Lakeside Recreation Areas

XXXXX

Stackpole Books

Lakeside Recreation Areas

Bill & Phyllis Thomas

Published by
STACKPOLE BOOKS
Cameron and Kelker Streets
P.O. Box 1831
Harrisburg, Pa. 17105

Published simultaneously in Don Mills, Ontario, Canada by Thomas Nelson & Sons, Ltd.

Printed in the U.S.A.

Library of Congress Cataloging in Publication Data

Thomas, Bill, 1934–
    Lakeside recreation areas.

    1. Lakes—United States—Recreational use—Directories. 2.
Reservoirs—United States—Recreational use—Directories. 3. Camp
sites, facilities, etc.—United States—Directories. 4. United States.
Army. Corps of Engineers—Civil functions. I. Thomas, Phyllis,
1935— joint author. II. Title
GV191.35.T47 1977   333.7'8'0973   76–54716
ISBN 0–8117–2104–3

Dedicated to
David, Dianne, Billy, Lisa,
and Alan,
who share our love for the water

CONTENTS

ACKNOWLEDGMENTS

The authors wish to offer sincere thanks to the many individuals and agencies—local, state, and federal—that provided information and photographs for this book, and especially to Francis X. Kelly of the U.S. Army Corps of Engineers, to Kampgrounds of America, Inc., Billings, MT, and to Bob Stewart and Bob Hudson of El Dorado Industries, Minneapolis, KS, manufacturers of recreational vehicles.

AUTHORS' NOTE

This volume is a guide to many of America's lakeside recreation areas constructed and maintained on U.S. Army Corps of Engineers impoundments. While we do not present this as a total list, it does include many of those which the authors feel rate high in the scope of quality recreational experiences which they offer. Readers may have encountered other such areas which are not included here, but had we endeavored to include all such facilities, this work would have covered several volumes. We hope you will use this book to search out your own outdoor experiences and that by doing so you will enjoy some of the best the outdoors has to offer.

INTRODUCTION

From nearly the beginning of time, water has held a vital attraction for mankind. It is no different today. Each year, more and more Americans seek out places to boat, water ski, swim, fish, canoe, and dive. When the Bureau of Outdoor Recreation issued its master survey for the United States more than 10 years ago, it revealed 75 percent of those people seeking outdoor recreation prefer to do it on or near water. Vacation activities for all ages and socioeconomic groups seem to be water-oriented.

It is no great surprise then that, following World War II, this latent desire began to surface. And with a flourishing economy that monthly saw our society become more and more affluent, the means to enjoy water recreation developed into reality. It was about this same time that Congress granted to the U.S. Army Corps of Engineers authority to "construct, maintain and operate public park and recreational facilities in reservoir areas and to permit the construction, maintenance and operation of such facilities."

The Corps has more than met its obligations to provide those facilities, despite the strong criticism it has incurred among environmentalists on some of its projects. Across the width and breadth of mainland America, the Corps has focused its recreation program on more than 400 major lakes involving more than 11 million acres of land and water with 40,000 miles of shoreline, enough to stretch from New York City to San Francisco 20 times.

Many of these man-made lakes are located in areas of the country as strikingly beautiful and scenic as any in the national parks or forests. It is small wonder that such reservoirs as Lake Sidney Lanier northeast of Atlanta have consistently attracted millions of fun-seekers year after year. In 1975, for instance, more than 13 million visitors came to Lake Sidney Lanier, a greater number than visited Yellowstone, Yosemite, and the Great Smoky Mountains National Park combined. Another 10 million enjoyed Lake Texoma on the Oklahoma-Texas border, the second most popular Corps recreation project.

It's readily apparent by now that the public desire for recreation is not going to diminish because of an energy shortage, nor is the attraction of water as a stage for outdoor recreation going to lessen. But even at such popular centers as Georgia's Lake Sidney Lanier on a holiday weekend, one seldom feels the urgency for additional elbow room. These lakes are vast, the facilities many. Counting areas leased to state and municipal governments or private concessionaires, there are more than 45,000 campsites, 50,000 picnic sites, and nearly 4,500 boat-launching ramps. Below the Corps of Engineers dams, there are canoeing and stream fishing. And you can also visit and see the mechanical operations of nearly all the dams which impound these reservoirs.

You may be surprised to learn that more than 800 recreation areas on Corps impoundments are managed by other federal, state, or local agencies. More than 200 fish and wildlife areas are leased to other federal agencies and states, areas often open for public use or environmental education. In these areas one may find such diverse outdoor recreation facilities as nature and hiking trails, golf courses, tennis courts, riding stables, restaurants, lodges, marinas, and even outdoor theaters.

In view of the incredibly varied recreational activities available at Corps reservoir sites, it is little wonder that they are so popular. With close to 350 million Americans enjoying the Corps' man-made lakes, a guide to direct more people to them would seem unnecessary. But the truth is that while many of these lakes enjoy great popularity, a good many others are relatively unknown to people outside their immediate areas. Even in the case of Lake Sidney Lanier, the majority of visitors come from within a radius of 100 miles, but they come again and again.

In this book, we have not attempted to chronicle every recreation area on every Corps of Engineers impoundment. Instead we have attempted to provide you a detailed account of what we feel to be some of the major areas—101 in all—to give you some direction and insight, not only about how to reach these areas, but what you find there once you arrive. At the end of each state section, the names and locations of other Corps man-made lakes are given.

Some things to keep in mind as you plan your visit to Corps lakes are these:

FREE INFORMATION: The Corps of Engineers will be happy to provide free literature and information about each of its projects. In addition to pamphlets on individual projects, the Corps publishes regional brochures for the West, Southwest, Midwest, Southeast, Northeast, and New England. The name and address of the Corps' district office administering each major lakeside recreation area discussed in this book will be found at the end of the section devoted to that particular recreation area.

It is always advisable to write in advance for information about the reservoir(s) you plan to visit. Most of these lakes have more than one recreation area, and each may offer different facilities. If you're not making prior plans, it would be a good idea to stop for information at the Corps of Engineers visitor centers located near the dams at many of these projects.

For information on nearby restaurants, lodging accommodations, and additional area recreational opportunities, write to the chambers of commerce in the nearest cities or towns.

FEES: There are no entrance fees to any Corps of Engineers recreational facilities. A camping fee is often charged, but at every reservoir where there are Corps-operated charge camping areas, the Corps' policy is to provide one area free. These free camping areas may be primitive in nature, but will contain designated campsites, sanitary facilities, and access roads. All Corps-managed boat ramps are free of charge except for mechanical lifts. A few recreation areas collect fees for group use. All other Corps-operated facilities are free of charge, though recreation areas operated by state or local governments or private enterprises on Corps lakes may charge fees for services or use of facilities. *But at all projects there are free areas where recreation activities may be enjoyed without cost.*

RESERVATIONS AND TIME LIMITS: No reservations are necessary except for group camps. Most campsites have a 14-day limit, especially during heavy-use periods.

HUNTING AND FISHING: Except where noted in the text dealing with specific recreation areas, all hunting and fishing at Corps of Engineers projects must be in accordance with the laws of the state in which these reservoirs are located. Hunting is permitted only in designated areas.

SEASON FOR USE OF RESERVOIRS: Special efforts are made to provide year-round recreation use, where project location and budget permit.

WATER LEVEL: It is important to remember that most Corps of Engineers reservoirs are for multipurpose use, and the water level at some may fluctuate drastically at certain times of the year, or during flood or drought.

Keep in mind that additional projects and facilities are being added constantly; so you may find more than are listed in this publication at your favorite lake or chosen destination.

If you've never discovered one of the Corps of Engineers' lakeside recreation areas for yourself, it's high time you study this book carefully, select a destination, and be off on a journey to unforgettable adventure.

STATE-BY-STATE GUIDE

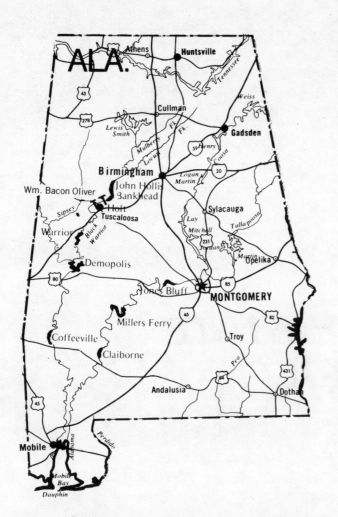

ALABAMA

CLAIBORNE LAKE

Lying in the gently rolling lowlands of the coastal plain in Alabama, Claiborne Lake is the southernmost impoundment on the Alabama River as it flows toward a convergence with the Tombigbee River. Together the two streams form the mighty Mobile River for a final run to the Gulf of Mexico approximately 80 river miles to the south of Claiborne Lock and Dam. A narrow reservoir, Claiborne Lake remains largely within the original riverbanks for its entire 60-mile length. Though open to navigation since 1969, recreational facilities have just recently been completed. Visitors will find a full range of water-based activities, with campsites along the lakeshore and motels available in nearby communities.

HOW TO GET THERE: From Monroeville, take SR 41 north for nearly 10 mi. to County Rd. 17. Turn northwest about 8 mi. to Wainright; then turn west to the dam road near Finchburg.

FISHING: Fishing is Claiborne Lake's main attraction. Primary species are largemouth bass, bream, and channel catfish. Nearly a dozen boat launching ramps; rental boats and supplies nearby. No marina at present, but one should be open soon.

HUNTING: Great quail and dove hunting here. Hunters will also find deer, wild turkey, squirrel, rabbit, raccoon and opossum. Hunting allowed on designated reservoir lands and in several state wildlife management areas less than an hour's drive away.

CAMPING: Both primitive and developed campsites at Corps of Engineers recreation areas, though the number at present is limited. More sites are being added all the time. Facilities include restrooms, drinking water, tables, grills, and launching ramps. Additional lakeshore campsites available at Riverview Camperama, a private enterprise, and at Little River State Park and Forest some 30 miles to the southeast.

OTHER ACTIVITIES: Swimming, water skiing, skin and scuba diving, boating; rental boats nearby. A lakeside marina is planned in the near future. Nearly a dozen Corps of Engineers launching ramps at present. Modern-day Huck Finns might enjoy locking through at Claiborne Dam and following the river system to Mobile and its expansive bay. Picnicking and hiking around reservoir. County Rd. 49, about 10 mi. north of Claiborne Dam, leads to a free ferry across the reservoir.

AREA ATTRACTIONS: Nearby Monroeville is noted as the home of novelist Harper Lee; here she, her brother, and a friend named Truman Capote garnered

the childhood experiences which were the basis for her Pulitzer Prize-winning book, *To Kill a Mockingbird.* The Old Fort Claiborne ruins, fossil beds, and a trail marking the explorations of the Spanish explorer de Soto are near the reservoir. At the Hank Williams birthplace and memorial near Evergreen, you can see mementos of the country and western singer's life. The Mobile Delta, second largest river delta in the United States, extends southward from the meeting place of the Alabama and Tombigbee rivers. In the upper Delta visitors will find scenery typical of the type many people associate with the Deep South—bayous, swamps, and hardwoods draped in Spanish moss.

FOR ADDITIONAL INFORMATION:

Public Affairs Officer
U.S. Army Engineer District, Mobile
P.O. Box 2288
Mobile, AL 36328

Resource Manager
Alabama River Lakes
P.O. Box 418
Camden, AL 36726

Park Manager
Little River State Park
Route 2, Box 77
Atmore, AL 36502

HOLT LAKE

Small but lovely Holt Lake extends northeast from Tuscaloosa along the Black Warrior River, one link in a series of locks, dams, and reservoirs which comprise the navigational system of the Black Warrior–Tombigbee Waterway. Tree-covered hills guard the 45-mile shoreline of this narrow Alabama lake, which backs up behind Holt Lock and Dam for 18.5 miles. A water surface of 3,200 acres provides recreation for the throngs of annual visitors who make this waterway one of the Corps of Engineers' 20 most popular areas. Campsites on the lakeshore are augmented by the most modern of facilities in nearby cities.

HOW TO GET THERE: From Tuscaloosa, head north 10 mi. on SR 116, then turn left on Lock and Dam Rd.

FISHING: Largemouth, smallmouth, white, and spotted bass vie with crappie as anglers' favorites. Also bream and catfish. Several launching ramps.

HUNTING: On designated reservoir lands and in three nearby state wildlife management areas. Quail hunting is tops. Also deer, wild turkey, dove, squirrel, rabbit, raccoon, and opossum.

CAMPING: Nearly 100 developed Corps of Engineers campsites; facilities include restrooms, drinking water, tables, grills, and launching ramps. Numerous private campgrounds around Tuscaloosa and Birmingham.

OTHER ACTIVITIES: Holt Lake is noted for its boating and water skiing, but swimming is also popular. Skin and scuba diving permitted. Several boat launching ramps. Canoeing on Locust Fork north of Birmingham. Picnicking and hiking trails. Additional

Lakefront campsites are often rustic and private. (TVA photo)

recreational opportunities in Talladega National Forest to the south.

AREA ATTRACTIONS: Mound State Monument near Moundville has 40 ceremonial mounds and a museum. The campus of the University of Alabama at Tuscaloosa offers a Museum of Natural History. Facilities at Oak Mountain State Park near Pelham include tennis, golf, horseback riding, roller skating, bicycling, and a unique demonstration farm area. Birmingham offers a number of attractions, but chief among them is Vulcan, a huge iron statue atop a mountain overlooking the city (elevators take visitors to the top of the giant figure). At Carrollton you can visit Ma Cille's Museum of Miscellania with its unique background. It's a one-woman museum created as a labor of love by a woman who, as a child, was too poor to go anywhere and wanted to give her children and others a place where "they could go free." While in town, investigate the story of the face in the courthouse window.

FOR ADDITIONAL INFORMATION:

Public Affairs Officer
U.S. Army Engineer District, Mobile
P.O. Box 2288
Mobile, AL 36628

Resource Manager
Holt Lake
P.O. Box 520
Demopolis, AL 36732

Forest Supervisor, Talladega
National Forests in Alabama
P.O. Box 40
Montgomery, AL 36101

LAKE DEMOPOLIS

A narrow, serpentine lake with so many twists and bends one might get dizzy following them, Demopolis Lake nearly surrounds the west central Alabama town from which it takes its name. Demopolis Dam just west of town creates a 10,000-acre reservoir which backs up through the white bluffs of the Tombigbee River to its confluence with the Black Warrior River four miles away. From there the huge lake extends 68 miles up the Tombigbee and 53 miles up the Black Warrior, forming 500 miles of shoreline along its irregular path. The six reservoirs of the Black Warrior—Tombigbee Waterway (Demopolis Lake among them) comprise one of the most popular Corps of Engineers recreation areas in the country, annually attracting nearly 4 million visitors.

HOW TO GET THERE: From Demopolis, head 6 mi. west on U.S. 80 to Lock and Dam Rd.

FISHING: Sportsmen in the know claim Demopolis Lake is unexcelled for bass, bream, and crappie. Numerous creeks, coves, and inlets provide plenty of getaway spots for the boating fisherman. Nearly a dozen Corps of Engineers launching ramps; supplies and rental boats nearby.

HUNTING: The Demopolis Waterfowl Management Area adjoins the lake. Duck hunting is very popular. Other game includes deer, wild turkey, squirrel, rabbit, quail, dove, raccoon, and opossum.

CAMPING: Nearly 150 Corps of Engineers campsites, both primitive and developed. Facilities include restrooms, picnic tables and grills, drinking water, dump station, and launching ramps. Talladega National Forest to the northeast and Chickasaw State Park to the south offer additional campsites. More campgrounds are being developed in the area.

OTHER ACTIVITIES: Water skiing, boating; rental boats and supplies nearby. Swimming at Payne Lake beach in Talladega National Forest; a wading pool for children at Chickasaw State Park. Picnicking along the shoreline of Lake Demopolis. Hiking trails in Talladega National Forest and Chickasaw State Park.

AREA ATTRACTIONS: Many antebellum homes in the nearby area may be toured; Gainswood at Demopolis is considered the most magnificent mansion in the Yellowhammer State. Choctaw National Wildlife Refuge near Coffeeville adjoins Coffeeville Lake, another Corps of Engineers reservoir. There's a federal fish hatchery at Marion which may be visited. Forty ceremonial mounds mark the site of an ancient city at Mound State Monument and Museum in Moundville. A tour of the University of Alabama campus at Tuscaloosa should include a visit to its Museum of Natural History, where the Hodges meteorite, the only one known to have struck a human being, is on display.

FOR ADDITIONAL INFORMATION:

Public Affairs Officer
U.S. Army Engineer District, Mobile
P.O. Box 2288
Mobile, AL 36328

Resource Manager
Demopolis Lake
P.O. Box 520
Demopolis, AL 36732

Forest Supervisor
National Forests in Alabama
P.O. Box 40
Montgomery, AL 36101

MILLERS FERRY RESERVOIR

A vast reservoir which annually attracts nearly 2 million visitors, Millers Ferry Reservoir lies amid the gentle hills and broad valleys of Alabama's coastal plain. This 17,000-acre lake follows the course of the Alabama River as it twists its way through the deep bends of the channel near Camden. Numerous fingers of water invade tributary creeks along the way, creating over 500 miles of irregular, tree-dotted shoreline. For 103 miles the lake backs up behind Millers Ferry Lock and Dam; and though it was officially designated William ''Bill'' Dannelly Reservoir by Congress to honor a local judge, it appears as Millers Ferry Reservoir on many maps. Corps of Engineers recreational facilities on both sides of the lake, as well as a state park at the water's edge, serve the needs of visitors.

HOW TO GET THERE: From Selma, SR 22 and SR 41 south both lead to recreation areas.

FISHING: Prime species are largemouth bass, bream, and channel catfish; and fishing is so good that 86 percent of those who visit here try their hand at it.

Lakeside marinas, rental boats, supplies; over a dozen launching ramps.

HUNTING: In season on designated reservoir lands and two state wildlife management areas, each less than an hour's drive from Camden. A bumper crop of

dove and some of the nation's best quail hunting are found here. Other species include deer, wild turkey, squirrel, rabbit, raccoon, and opossum.

CAMPING: Corps of Engineers campsites are limited at present (around 25, both primitive and developed), but plans call for more to be added to the system as they are completed. Facilities include restrooms, drinking water, showers, tables, grills, and launching ramps. Camden State Park has nearly 50 campsites with trailer hookups and a dump station, as well as vacation cottages. Nearby Talladega National Forest offers additional campsites.

OTHER ACTIVITIES: Swimming, water skiing, pleasure boating; over a dozen launching ramps, lakeside marinas, rental boats. Canoeing on the Cahaba River. Adventurous scuba divers may dive for gold at Possum Bend near Camden, where a riverboat went down with a reported quarter-of-a-million aboard. Try your hand at on-land treasure hunting, too; this area abounds with prehistoric Indian and Civil War relics. Numerous picnic areas and hiking trails. Camden State Park offers bicycling and a nine-hole golf course,

while Talladega National Forest to the north provides additional recreational opportunities.

AREA ATTRACTIONS: At Selma, you can tour Sturdivant Hall, a restored antebellum mansion complete with its own ghost. The nearby ruins of Cahaba are just that—the ruins of the city originally chosen as Alabama's state capital. At Marion you can visit a federal warm-water fish hatchery.

FOR ADDITIONAL INFORMATION:
Public Affairs Officer
U.S. Army Engineer District, Mobile
P.O. Box 2288
Mobile, AL 36628

Resource Manager, Dannelly
Alabama River Lakes
U.S. Army Corps of Engineers
P.O. Box 418
Camden, AL 36726

Camden State Park
P.O. Box 128
Camden, AL 36726

Forest Supervisor, Talladega
National Forests in Alabama
P.O. Box 40
Montgomery, AL 36101

ADDITIONAL CORPS OF ENGINEERS LAKES*

Coffeeville Lake (from Coffeeville, 3.5 mi. W on U.S. 84); Jones Bluff Lake†, ‡, § (from Montgomery, U.S. 80 W to dam at Benton); Warrior Lake (U.S. 43 to Eutaw, SR 14 S to Lock and Dam Rd.)

ARIZONA

ALAMO LAKE

A unique desert recreational area in west central Arizona, Alamo Lake is fed by the Bill Williams River as it makes its way to a meeting with the Colorado River some 40 miles away. On every side, the sands march away to meet the mountain ranges which surround this area. The river and a nearby mountain range are named for one of the colorful mountain men who roamed the Southwest in the nineteenth century. Alamo Lake State Park, built around the shoreline, is one of the few known regions in the world where the Joshua tree and saguaro cactus grow side by side. Reservoir land adjoining the state park is managed by the Arizona Department of Fish and Game as a wildlife preserve.

---

*Unless designated as follows, each project has restrooms, drinking water, and developed campsites. † = no restrooms; ‡ = no drinking water; § = no developed campsites.

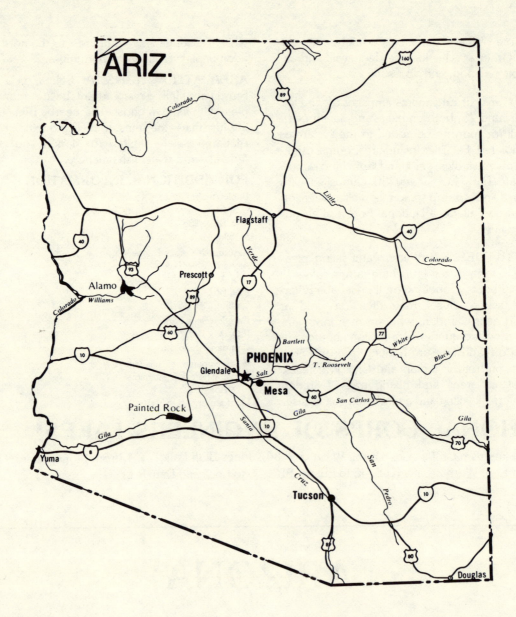

HOW TO GET THERE: Follow state park road north of Wenden for approximately 40 mi. Wenden is on U.S. 60.

FISHING: Rainbow and cutthroat trout, as well as largemouth bass, tempt anglers on this 500-acre lake. Two launching ramps.

HUNTING: White-tailed deer, dove, quail, and duck are found around the reservoir. Javelina (wild desert pigs) and desert bighorn sheep have also been sighted in the area.

CAMPING: Nearly 50 campsites, both primitive and developed, in Alamo Lake State Park; facilities include restrooms, showers, drinking water, tables, grills, full hookups, and launching ramps. Private campgrounds within a half-hour's drive.

OTHER ACTIVITIES: Swimming, water skiing, boating; launching ramps. Picnicking, rockhounding, birdwatching, and nature study are favorite on-land pastimes.

AREA ATTRACTIONS: Lake Havasu National Wildlife Refuge near Needles, CA, offers a look at native wildlife residents, including beaver, bighorn sheep, wild burro, the rare Harris hawk, and migratory waterfowl. Wild and remote Palm Canyon near Quartzsite is the only place in the state where wild palms are found, but visitors must do some arduous climbing to see them. In Quartzsite's old cemetery, the unusual Hi Jolly Memorial has been erected to one of the men who helped introduce camels as a means of transportation in Arizona. The Colorado River Indian Reservation is near Parker. On Burro Creek, southeast of

Wikieup, is a desert canyon vista; despite intense desert heat, this canyon holds water the year around.

FOR ADDITIONAL INFORMATION:
Public Affairs Officer
U.S. Army Engineer District, Los Angeles
P.O. Box 2711
Los Angeles, CA 90053

ADDITIONAL CORPS OF ENGINEERS LAKES

Painted Rock Dam (near Gila Bend).*

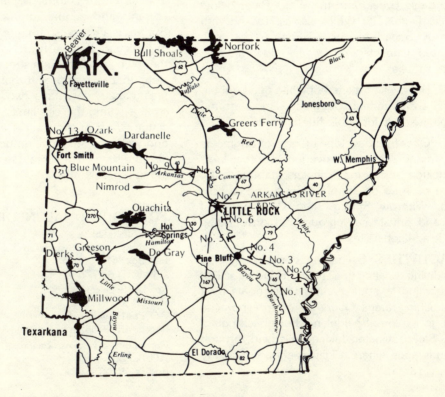

ARKANSAS

BEAVER LAKE

An offspring of the White River, majestic Beaver Lake forms a 70-mile-long waterway through a series of scenic valleys in the northwestern corner of Arkansas. Huge limestone bluffs, varying in

---

*This project has restrooms and drinking water but no developed campsites.

height and color in each arm of the reservoir, form a rugged boundary at the water's edge. The countryside is unspoiled and the sparkling, clear water unpolluted. Fall color in this rustic mountain region, covered with lush hardwoods, is outstanding. Visitors will find every type of accommodation, ranging from campgrounds to luxury resorts, all around the lake.

HOW TO GET THERE: From Fayetteville, take U.S. 71 north for 22 mi.

FISHING: The main attraction at Beaver Lake. Noted for its largemouth and smallmouth bass, the reservoir also produces crappie, bream, channel catfish, northern pike, walleye, and white bass. Three excellent smallmouth streams are in the area—Kings River, War Eagle Creek, and the White River below Beaver Lake Dam. The White also is stocked with rainbow trout. Spearfishing in reservoir is seasonal. Over a dozen launching ramps; lakeside marinas, rental boats, supplies.

HUNTING: Deer, wild turkey, bobwhite quail, rabbit, and squirrel are abundant. Designated areas around the reservoir are open for public hunting.

CAMPING: Over 500 developed Corps of Engineers campsites, many on bluffs overlooking the lake; facilities include restrooms, showers, tables, grills, drinking water, firewood, dump stations, and launching ramps. Withrow Spring State Park north of Huntsville and private campgrounds in surrounding area provide additional sites.

OTHER ACTIVITIES: Swimming, water skiing, skin and scuba diving, boating of all types, including sailing and houseboating; lakeside marinas, rental boats, over a dozen launching ramps. Over 450 miles of shoreline for boaters to explore. Spelunking, horseback riding, picnicking. Steep, winding hiking trails lead through forested landscapes, where wildlife is abundant. Many scenic drives in area.

AREA ATTRACTIONS: Visitors may take a self-guided tour of the Civil War battlefield where the Union secured Missouri at Pea Ridge National Military Park near Rogers. The Daisy Air Rifle plant at Rogers offers tours and the International Air Gun Museum. At Fayetteville, the campus and many of the buildings of the University of Arkansas may be toured. Tontitown, a small Italian colony, is famous for its unusual restaurants and annual summer grape festival. Eureka Springs is a village so unique that it has been featured in Ripley's "Believe It or Not." Homes viewed from the town's one through street cling to valley slopes so steep that they appear to have just one story when in reality they may have two or three. There are 63 springs within the city limits. Nearby the magnificent statue *Christ of the Ozarks* towers seven stories high and weighs over 1 million pounds. Eureka Springs is also the site of The Great Passion Play, presented during summer months.

FOR ADDITIONAL INFORMATION:
Public Affairs Officer
U.S. Army Engineer District, Little Rock
P.O. Box 867
Little Rock, AR 72203

Resident Engineer
Beaver Lake Office
P.O. Box Drawer H
Rogers, AR 72756

Greater Beaver Lake Association
P.O. Box 308
Rogers, AR 72756

BULL SHOALS LAKE

Sprawling like a giant serpent across three north central Arkansas counties, with fingers stretching into Missouri, Bull Shoals Lake creates 1,000 miles of rugged shoreline in the scenic Ozark Mountains. Over 20 public recreation areas, as well as a wide range of private accommodations ranging from rustic to luxurious, are located along the lakeshore and on the banks of the White River, which flows south from Bull Shoals Dam. The Ozarks are noted as one of the top seven pollution-free areas of the nation, and the lake's clear, sparkling water attests to that fact. A vast variety of year-round recreation opportunities, both on and near the water, make this one of the 20 most popular Corps of Engineers reservoirs in the country.

Most lakes offer full-service marina facilities. (Arkansas Dept. of Parks and Tourism photo)

HOW TO GET THERE: From Mountain Home, AR, take SR 5 north, then SR 178 west for a total of 17 mi. to the village of Bull Shoals. From Branson, MO, take U.S. 160 east for 70 mi. or less; the lake is just a few miles south of this highway over several access roads.

FISHING: Bass, crappie, channel cat, bream, and walleye all abound in Bull Shoals Lake, but it's the lunker bass which have gained national acclaim for the reservoir, with catches to 12 pounds not uncommon. As of spring, 1975, Bull Shoals had produced Ar- kansas state record smallmouth, largemouth, and spotted bass, as well as the Missouri state record large- mouth. White River below the dam has a well- justified reputation for its rainbow and brown trout fishing (state records for both came from the White), as well as smallmouth bass. Rugged Crooked Creek, about 15 miles south of the town of Bull Shoals, is considered Arkansas's blue-ribbon smallmouth stream. A little over 30 miles south of the same town is the magnificent Buffalo River, a free-flowing stream also noted for its smallmouth fishing. Seasonal spearfishing allowed in reservoir. Over 20 launching ramps; lakeside marinas, rental boats, supplies.

HUNTING: Deer, squirrel, rabbit, and other small game animals are plentiful. Also found here are duck, quail, and wild turkey.

CAMPING: Over 625 developed Corps of Engineers campsites; facilities include restrooms, drinking water, showers, tables, grills, firewood, electrical hook-ups, dump station, and launching ramps. Boat camping is permitted both along the 1,000-mile lakeshore and on the banks of the White River. In addition, Bull Shoals State Park (AR) offers lakeside and riverside campsites; and the city of Forsyth (MO) operates a campground on the lake. Numerous private campgrounds in area.

OTHER ACTIVITIES: Swimming, skin and scuba diving, water skiing, boating; lakeside marinas, rental boats (including sailboats and houseboats), over 20 launching ramps. Float trips and canoeing on the White and Buffalo rivers, as well as Crooked Creek (a guide is recommended on this stream). Picnicking, horseback riding, and spelunking are favorite area activities. Hiking trails and mountain drives take visitors to scenic views.

AREA ATTRACTIONS: Mountain Village 1890 is an authentically restored Ozark town near the village of Bull Shoals. Adjacent Bull Shoals Caverns contain an interesting array of rock formations. Not far from here atop Bull Mountain is Top o' the Ozark, an observation tower which affords a panoramic view of the Ozark hills. Many seasonal festivals in the area feature Ozark mountain music. A free ferry crosses the lake near its center (take SR 125 north of Peel for five miles). The Buffalo National River Park, with headquarters at Harrison, offers spectacular scenery.

FOR ADDITIONAL INFORMATION:

Public Affairs Officer
U.S. Army Engineer District, Little Rock
P.O. Box 867
Little Rock, AR 72203

Resident Engineer
Bull Shoals Lake
Mountain Home Resident Office
Mountain Home, AR 72653

Bull Shoals Lake and White River Association
727 S. Bryant
Mountain Home, AR 72653

Superintendent
Bull Shoals State Park
Bull Shoals, AR 72619

De Gray State Park's modern lodge and convention center for luxurious "roughing-it." (Arkansas Dept. of Parks and Tourism photo)

De GRAY LAKE

Nestled in the Caddo River Valley in the rolling foothills at the southern extremity of the Ouachita Mountain range, sprawling De Gray Lake offers over 200 miles of shoreline and 13,400 acres of deep blue water. Located approximately 20 miles southwest of Hot Springs, AR, the reservoir is one of the newest Corps of Engineers projects in the Southeast. Some of the most sophisticated public use areas yet developed by the Corps are at De Gray Lake, and development is not yet complete. The scenic countryside is forested with shortleaf and loblolly pine, as well as various hardwoods. Ouachita National Forest to the north offers additional recreational opportunities. Lakeshore accommodations range from wilderness campsites to one of the most luxurious state park lodges in the country. De Gray State Park Lodge and Convention Center occupies a drive-to island on the reservoir. Opened in November, 1975, it is part of a $6 million recreation complex, which also includes vacation cottages.

HOW TO GET THERE: Approximately 20 mi. southwest of Hot Springs, via SR 7.

FISHING: Over 8,000,000 fish have been stocked here. Black bass, bluegill, crappie, goggle-eye perch, shad, and catfish are supplemented with exotic game fish such as northern pike, muskellunge, and walleye. A state record pike was caught here. The Caddo River is a good smallmouth stream; it can be floated from Norman to De Gray Reservoir. Spearfishing in the lake is seasonal. Nearly 20 launching ramps; lakeside marinas, rental boats, supplies.

HUNTING: Game species on the project include quail, deer, wild turkey, squirrel, rabbit, and duck.

CAMPING: Over 250 Corps of Engineers campsites, both primitive and developed. Facilities include restrooms, showers, drinking water, tables, grills, electrical hookups, dump stations, and launching ramps. Wilderness camping is available at five areas around the lake. Additional sites with full hookups in De Gray State Park, a state recreation area opened in late 1975.

OTHER ACTIVITIES: Swimming, water skiing, boating; lakeside marinas, rental boats, nearly 20 launching ramps. The Caddo River is a floatable stream for 25 to 30 miles above De Gray Lake. Bird-watchers may see blue heron, osprey and, during the winter, bald eagle. A variety of wildlife includes such nongame species as beaver and armadillo. Numerous picnic areas in scenic settings dot the shoreline. A trail system, when completed, will include 40 miles of hiking, nature, bicycle, and motorcycle trails. In November, 1975, a $6 million complex was opened at De Gray State Park. Built and operated by the state of Arkansas, the recreation area offers a full range of activities, such as horseback riding, tennis, and golf on a 7,200-yard championship PGA course. Opportunities for backpacking and nature study abound in Ouachita National Forest 30 miles north of the reservoir.

AREA ATTRACTIONS: One of the most popular attractions in the Razorback State and probably the best-known mineral spa in the nation is Hot Springs National Park, about 20 miles north of De Gray Lake. Dryden Pottery in the city of Hot Springs offers guided tours to the public. Crater of Diamonds State Park near Murfreesboro is an area unique in this country; visitors may search for real diamonds and keep any they find. The Caddo Indian Burial Mounds and an adjoining museum are open to the public.

FOR ADDITIONAL INFORMATION:
Public Affairs Officer
U.S. Army Engineer District, Vicksburg
P.O. Box 60
Vicksburg, MS 39180

Lake Manager
De Gray Lake
P.O. Box 497
Arkadelphia, AR 71923

The complete resort facilities at De Gray State Park include this 96-room lodge. (Arkansas Dept. of Parks and Tourism photo)

GREERS FERRY LAKE

Nestled in the eastern foothills of the Arkansas Ozarks at the foot of beautiful Pryor Mountain, Greers Ferry Lake is in the heart of one of middle America's foremost recreational areas. Although most of the reservoir's 300-mile shoreline remains undeveloped, Greers Ferry Lake offers something to everyone, from wilderness to the utmost in luxury. Over 2.5 million visitors come here annually to sample the wide range of recreational opportunities. One of the loveliest bodies of water in a state known for its magnificent lakes, Greers Ferry Lake backs up the Little Red River Valley some 50 miles, spilling into countless draws in the hills to form secluded coves and inlets and creating a water surface of 40,000 acres. The Corps of Engineers and local businessmen jointly sponsor an annual lakewide cleanup that has received national recognition, assuring visitors one of the most litter-free impoundments in the United States.

HOW TO GET THERE: From Little Rock, go 15 mi. north on U.S. 67-167, then 50 mi. north on SR 5.

FISHING: Every game fish native to the state of Arkansas has been stocked in Greers Ferry. White bass and walleye head up feeder streams during the spring spawning runs, providing good catches for both boat and bank fishermen. Some of the most predictable good fishing occurs during the crappie spawn later in the spring. Largemouth and striped bass, chain pickerel, bream, and catfish also provide plenty of action for lake anglers. Rainbow trout is stocked in the lake and in tailwaters below the dam. Some smallmouth bass are found in the Little Red River, with the South Fork probably the most productive and the Middle Fork best for float fishing. Record catches are common occurrences, and major bass tournaments are held here frequently. In early summer, night fishing with lights for crappie, white bass, and trout is a popular pastime. Seasonal spearfishing permitted in lake. Over 20 launching ramps; lakeside marinas, rental boats, supplies.

HUNTING: Deer, quail, ducks, geese, squirrels, rabbits, and wild turkey are ample in the pineclad hills which surround the lake. The 10,000-acre Gulf Mountain State Wildlife Management Area lies within a 15-minute drive of the west end of the reservoir. Hunting also allowed in designated areas along lakeshore.

CAMPING: Around 1,000 Corps of Engineers campsites, all developed, with more being added all the time. Facilities include restrooms, drinking water, showers, tables, grills, dump stations, electrical hookups, and launching ramps. Plenty of private campgrounds near the lake and on the banks of the Little Red River.

OTHER ACTIVITIES: Swimmers will find this reservoir to be one of the most pollution-free lakes in the country, while skin and scuba divers are lured here by the clear, deep waters. Water skiing in plenty of wide-open spaces. Every type of boating imaginable is done here; lakeside marinas rent ski boats, fishing boats, party barges, paddle boats, pontoon boats, sailboats, houseboats, and canoes. Over 20 launching ramps. Lake cruises available. The Little Red River, both above and below the lake, provides some fine waters for float trips. Scenic picnic areas throughout the wooded hills which embrace the lake's shoreline. Hiking trails branch off in every direction. The outstanding Sugar Loaf Mountain Island Nature Trail, constructed by the Corps of Engineers, is part of the National Trails System, but you can reach it only by boat. It's located on a 250-acre island formed when the waters of Greers Ferry Lake inundated the mountain's base. A water-taxi service is available for non-boaters. Those who reach the top, 556 feet above the lake, are rewarded with a sweeping 360-degree view of this area's rugged beauty. Other activities available in towns and recreation areas surrounding the reservoir include bike and horseback riding, golf, tennis, pool swimming, archery, volleyball, roller skating, and bowling.

AREA ATTRACTIONS: A federal fish hatchery below Greers Ferry Dam permits visitors to see rainbow trout in various stages of growth. Sightseers will particularly enjoy the panoramic view from Millers Point on Pryor Mountain. The town of Heber Springs, near the Greers Ferry Dam site, boasts several mineral-water springs; and visitors are welcome to drink from them. Near Mountain View, approximately 35 miles to the north, lies the Ozark Folk Culture Center, an interpretive state park which displays the music, lore, arts, and crafts of this mountain region. Blanchard Springs Cavern, 15 miles northwest of Mountain View in the Ozark National Forest, is a relatively new cave find which will delight spelunkers.

FOR ADDITIONAL INFORMATION:
Public Affairs Officer
U.S. Army Engineer District
P.O. Box 867
Little Rock, AR 72203
Greers Ferry Lake and
Little Red River Association
P.O. Box 408
Heber Springs, AR 72543

LAKE DARDANELLE

The largest Corps of Engineers reservoir on the Arkansas River, Dardanelle stretches west of Russellville for 50 miles. Its 315 miles of picturesque shoreline are dotted with several towns and numerous recreational facilities, all easily accessible from a good highway system. Although Lake Dardanelle is one of Arkansas' most popular water playgrounds and attracts over 2 million visitors annually, it lies within an hour's drive of some of the state's most beautiful and scenic wilderness. Two national forests and three state parks, all within a 50-mile radius, provide additional recreational facilities.

HOW TO GET THERE: Just west of Russellville, via I-40.

FISHING: Largemouth bass get most of the attention in the reservoir, with fish in the eight- to nine-pound class caught surprisingly often in the stump-filled coves. The lake has also been stocked with bream, crappie, and striped bass (some 20-pounders have been reported). The Arkansas River below the dam yields white bass, as well as catfish which run in the 40-pounds-plus class. For pure sport, try the river's alligator gar, which have been confirmed at over 200 pounds. Big Piney Creek, which originates in the Ozark National Forest and flows south into Lake Dardanelle, is a popular smallmouth bass stream. Seasonal spearfishing in reservoir. Around 20 launching ramps; lakeside marina, rental boats, supplies.

HUNTING: Because the reservoir is near the Ozark and Ouachita national forests, an abundance of wildlife is found here. In-season hunting for deer, turkey, and wild boar, as well as squirrel, rabbit, ducks, quail, and dove. Hunting allowed in designated areas around lake, at Galla Creek State Wildlife Management Area 10 mi. southeast, and in both national forests.

CAMPING: Over 250 Corps of Engineers developed campsites; facilities include restrooms, drinking water, showers, tables, grills, dump stations, electrical hook-ups, and launching ramps. Excellent lakeside camping in Dardanelle State Park. Additional campsites available in Mount Nebo and Petit Jean state parks, as well as the Ozark National Forest—all within a few minutes' driving distance.

OTHER ACTIVITIES: Swimming in lake, water skiing, power boating and sailing; lakeside marinas, rental boats of all types, around 20 launching ramps. Public swimming pools at Russellville and Clarksville. Traveling through the locks and dams along the Arkansas River, boaters may cruise upstream to Tulsa, OK, or downstream to New Orleans. Houseboats and party boats are available here. Canoeing on nearby mountain streams is growing in popularity each year. Two of the area's wildest streams are Big Piney Creek and Illinois Bayou; they are not recommended for beginners in any season. Arkansas' Highway 7 heading north from Russellville to Harrison through the Ozark National Forest has been voted one of the country's 10 most scenic roads. For a spectacular view of forests, mountains, and lakes, drive to the top of Mount Magazine near Paris; it's the highest plateau between the Appalachians and the Rockies. Unlimited opportunities for hiking and backpacking in the country surrounding Lake Dardanelle. Scenic picnic areas. Tennis, basketball, and volleyball courts nearby, as well as several golf courses.

AREA ATTRACTIONS: Holla Bend National Wildlife Refuge near Russellville is a sanctuary for migrating waterfowl, and visitors are welcome. The late Winthrop Rockefeller's collection of antique autos is displayed in the Museum of Automobiles near Morrilton. Close by are Winrock Farms, open to the public, where prize-winning Santa Gertrudis cattle are bred in unique surroundings. Adjoining the vast farms is Petit Jean State Park, studded with natural scenic wonders. Wiederkehr Winery near Altus offers tours of its wine cellars and vineyards. The orchards near Clarksville, the peach capital of Arkansas, provide acres of blooms in the spring and fine eating later in the summer; an annual Peach Festival is held the weekend before the Fourth of July.

FOR ADDITIONAL INFORMATION:

Public Affairs Officer
U.S. Army Engineer District, Little Rock
P.O. Box 867
Little Rock, AR 72203

Dardanelle Lake Association
P.O. Box 987
Russellville, AR 72801

Forest Supervisor
Ozark National Forest
Box 340
Russellville, AR 72801

LAKE OUACHITA

Aptly nicknamed one of Arkansas' Diamond Lakes, Lake Ouachita offers 48,300 acres of incredibly blue water in a primeval setting of virgin pines and cedars. Located on the Ouachita River in the Ouachita Mountains of west central Arkansas, the vast reservoir reaches 52 miles upstream behind Blakely Mountain Dam and creates a shoreline 975 miles long. The haunting, pristine beauty of the area is preserved by the Ouachita National Forest, which completely surrounds the lake. Those who know Lake Ouachita best describe it in superlatives—largest lake completely within the state, one of the two most pure and unpolluted lakes in the United States, one of the two deepest lakes in midcontinent America, and one of the most beautiful bodies of water anywhere. Certainly no other region in the state can surpass this one when it comes to variety; there's something of interest to almost everyone. Accommodations on the lake range from luxury resorts to primitive island campsites, and numerous public and private facilities lie within a short driving distance. Hot Springs National Park, one of the nation's most famous spas, is located just a few miles away.

HOW TO GET THERE: From Hot Springs, travel west on SR 227 for 13 mi.

FISHING: Anglers rate the fishing here as excellent. Gamefish include largemouth, smallmouth, and spotted bass, white and black crappie, northern pike, muskellunge, bluegill, red ear sunfish, stripers, grass pickerel, walleye, and several species of bream. Trotliners can find channel cat, blue cat, and flatheads. White bass school in the spring and spawn upriver in very early spring. Trout thrive in the cold water around the dam, and there's a quality smallmouth area on the upper Ouachita River between Pine Ridge and Sims. Seasonal spearfishing in the lake. Over 15 launching ramps; lakeside marinas, rental boats, supplies.

HUNTING: "Ouachita" is the French spelling of an Indian word meaning "good hunting grounds." Game in the Ouachita National Forest includes deer, turkey, quail, dove, rabbit, squirrel, and waterfowl. Some of the best squirrel woods are along the small creeks in hardwood areas; and hunters also find wood duck here after the acorns have fallen in the autumn. Towards the last of the season, mallards usually fill up the South Fork Recreation Area. Wild turkey are taken in the Muddy Mountain Refuge and the area around Logan Gap Mountain. Both areas, as well as the Little Fir Recreation Area on the lake, are good for white-tailed deer.

CAMPING: Over 450 Corps of Engineers campsites, both primitive and developed, with more being added as public demand increases. Facilities include restrooms, showers, drinking water, tables, grills, and launching ramps. Lake Ouachita State Park near the dam offers sites with electrical hookups. For a true get-away-from-it-all wilderness camping experience, find a secluded spot on one of the more than 100 islands that dot the lake. A new dimension in camping is offered by Camp-A-Float cruisers, with headquarters at Crystal Springs Fishing Village. Campers with recreational vehicles can have them secured to a large pontoon boat, then take off to explore Lake Ouachita on their own floating campsite. (For more information, write Camp-A-Float, P.O. Box 1625, Rockford, IL 61110.) Additional campsites within Ouachita National Forest and at Hot Springs National Park near Hot Springs, AR. Several boat camping sites along the upper Ouachita River on U.S. Forest Service land.

OTHER ACTIVITIES: Swimming, skin and scuba diving, water skiing, boating of all types; lakeside marinas, all kinds of rental boats, over 15 launching ramps. Sightseeing boat trips available. Float trips and canoeing on the upper Ouachita River. Hikers, bicyclists, horseback riders, and motorists will find unsurpassed scenic beauty. Rockhounding is a rewarding pastime in an area abounding with quartz crystals, wavellite, and black quartz. Visitors may hunt and keep real diamonds at Craters of Diamonds State Park near Murfreesboro and Indian artifacts turn up throughout the area. Picnic areas are scattered all along the lakeshore and in the national forest. The Talimena Scenic Drive begins at Mena, AR, and winds across the mountaintops to Talihina, OK, on U.S. 271. Also near Mena on national forest land is the Caney Creek backcountry, designated an Eastern Wilderness Area in 1975, with its excellent backpacking opportunities. Over 120 miles of the Ouachita Trail, which will stretch for 195 miles through the heart of the Ouachita National Forest when completed, have been constructed and are open to the public.

AREA ATTRACTIONS: The popular resort city of Hot Springs features seasonal horse racing, night life, and Hot Springs National Park with its soothing thermal baths. Dryden Pottery in Hot Springs offers free guided tours. Wildwood 1884 in the same city is one of the city's oldest homes; it's open to the public. Children will enjoy Hot Springs' I.Q. Zoo, where trained animals appear in continuous shows. Queen Wilhelmina State Park near Mena lies atop one of the highest points between the Alleghenies and the Rockies; a miniature railroad follows a short track around the scenic mountaintop.

FOR ADDITIONAL INFORMATION:

Public Affairs Officer
U.S. Army Engineer District, Vicksburg
P.O. Box 60
Vicksburg, MS 39180

Suprintendent
Lake Ouachita State Park
Star Route
Mountain Pine, AR 71956

Forest Supervisor
Ouachita National Forest
Box 1270
Hot Springs, AR 71901

Park Superintendent
Hot Springs National Park
P.O. Box 1219
Hot Springs, AR 71901

Diamond Lakes Travel Association
P.O. Box 1500
Hot Springs, AR 71901

NIMROD LAKE

An impoundment of the Fourche La Fave River in west central Arkansas, Nimrod Lake forms part of the northern boundary of the vast Ouachita National Forest. Towering pines and hardwood trees cover the 77 miles of hilly shoreline. The lake lies in the Ouachita Mountain highlands, with the Ozark Mountains visible to the north. Visitors appreciate the serenity, beauty and variety of recreational facilities available year-round at this reservoir. Because it is off the beaten track in a state of many easily accessible lakes, Nimrod is seldom crowded. Three state parks, a national park, two national forests, one national and one state wildlife refuge—all within an hour's drive— provide additional opportunities for recreation. Facilities at nearby Petit Jean State Park include a lodge and cabins.

HOW TO GET THERE: From Hot Springs, head 40 mi. north on SR 7.

FISHING: Nimrod boasts some of the best crappie fishing in the state, in both the lake and river. Catfish grow large here. Fisherman also add largemouth bass, white bass, and bream to their stringers. Some restricted areas. A state goose sanctuary is closed to fishing between October 16 and April 1. Spearfishing allowed in season. Over half a dozen launching ramps; lakeside marina, rental boats, supplies.

HUNTING: Wild surroundings here provide good cover for a variety of game, and the lake is an ideal spot for a base camp. Wild turkey in nearby area. Over 80 food and cover plots developed for quail near the lake, as well as a duck area. Also deer, cottontail rabbit, gray and red squirrel, fox, raccoon, and opossum. Protected areas on the lake include a state wildlife refuge and a goose sanctuary.

CAMPING: Over 100 developed Corps of Engineers campsites; facilities include restrooms, drinking water, tables, grills, firewood, electrical hookups and launching ramps. Additional campsites in Ouachita National Forest. Other public and private campgrounds within a half-hour's drive.

OTHER ACTIVITIES: Many water skiers claim that Nimrod Lake is one of the best ski areas in the state. Swimming and boating also popular; lakeside marina, rental boats, over half a dozen launching ramps. Visi-tors who choose to explore this lovely hill country via numerous hiking trails will find beautiful vistas in every direction. Cove Mountain Trail and Long Hollow Trail, which parallel the lake on the south, provide a back-roads sightseeing experience for motorists. Picnic tables are scattered around the lakeshore.

AREA ATTRACTIONS: Nationally famous Hot Springs National Park, at Hot Springs, features hiking trails, mountaintop scenic drives and, of course, its numerous thermal springs, bath houses and hydro-therapy pools. The Holla Bend National Wildlife Refuge near Russellville welcomes visitors. Near Morrilton is the Museum of Automobiles, a showcase for the antique auto collection of the late Winthrop Rockefeller. At nearby Winrock Farms, Rockefeller made his home and employed some unique methods to raise his prize-winning Santa Gertrudis cattle; the farms are open to the public. Petit Jean State Park in the same area is the state's oldest and most popular park; it sits atop Petit Jean Mountain and offers natural scenic wonders.

FOR ADDITIONAL INFORMATION:

Public Affairs Officer
U.S. Army Engineer District, Little Rock
P.O. Box 867
Little Rock, AR 72203

Forest Supervisor
Ouachita National Forest
Box 1270
Hot Springs, AR 71901

NORFORK LAKE

Resembling a Chinese dragon, Norfork Lake snakes its way across the Missouri border into north central Arkansas. Five hundred and fifty miles of rocky shoreline twist and turn to create scores of coves and inlets. An impoundment of the North Fork River, this popular reservoir nestled in the Ozark Mountains provides a wealth of outdoor activities. The clear unpolluted water, scenic backdrop, fresh mountain air, and mild winters are an unbeatable combination for year-round fun. Many commercial establishments in the nearby area provide services and recreation facilities of all types.

HOW TO GET THERE: From Mountain Home, AR, take U.S. 62 north and east for 10 mi.; from Branson, MO, take U.S. 160 east for 60 mi.

FISHING: Rainbow trout which can run into the 15-pound class have earned a reputation for the North Fork River below the dam and nearby White River.

Nearly all varieties of freshwater game fish are found in the lake, including bass, walleye, crappie, bream, and catfish; but the crappie and lunker bass attract the most anglers. Night fishing with lights for white bass and crappie has become a Norfork tradition over the years. Spearfishing in season. The Buffalo National

Lakes on flyways provide great in-season duck hunting.

River nearby provides excellent float fishing for small-mouth bass. Over 25 launching ramps; lakeside marinas, rental boats, supplies.

HUNTING: Timbered hills provide good cover for white-tailed deer. Also hunters will find wild turkey, squirrels, rabbits, ducks, geese, doves, and quail.

CAMPING: Over 450 developed Corps of Engineers campsites; facilities include restrooms, drinking water, showers, tables, grills, dump stations, and launching ramps. Numerous private campgrounds; limited number of sites in the Sylamore District of the Ozark National Forest south of the reservoir. Boat camping is popular along the White River.

OTHER ACTIVITIES: Swimming, water skiing, skin and scuba diving, boating; over 25 launching ramps, lakeside marinas, rental boats. Sailing is particularly popular. Float trips and canoeing on the North Fork, White, and Buffalo rivers. Hiking trails and picnic areas (some group shelters). Two state-operated free ferries cross the lake, connecting U.S. 62 and SR 101.

AREA ATTRACTIONS: A federal trout hatchery below Norfork Dam welcomes vistors. The Arkansas State Wildlife Refuge borders part of the lake's western shore in Baxter County, Arkansas. Spelunkers will enjoy Blanchard Springs Caverns in nearby Ozark National Forest. Some 40 miles southeast of the lake lies the Ozark Folk Culture Center, an interpretive state park featuring folk music, crafts, and native folklore every day April through October and on winter weekends. Boggy Creek, which flows into an arm of the reservoir near Elizabeth, AR, abounds in legends of a mysterious, apelike creature (believed to be a relative of the Abominable Snowman of the Himalayas and Bigfoot of the Pacific Northwest) which inhabits the surrounding terrain.

FOR ADDITIONAL INFORMATION:
Public Affairs Officer
U.S. Army Engineer District, Little Rock
P.O. Box 867
Little Rock, AR 72203

Resident Engineer, Norfork Lake
Mountain Home Resident Office
Mountain Home, AR 72653

Norfork Lake Association
P.O. Box 3044
Mountain Home, AR 72653

ADDITIONAL CORPS OF ENGINEERS LAKES*

Blue Mountain Lake (from Little Rock, 68 mi. W on I-40, 22 mi. S on SR 7, 17mi. W on SR 10); Dierks Lake‡ (from Texarkana, 46 mi. N on U.S. 71, 11 mi. E on U.S. 70 to access road); Lake Greeson (from Little Rock, 60 mi. S on I-30 to SR 26, 37 mi. W to Murfreesboro, 6 mi. N on SR 19); Millwood Lake (From Texarkana, 19 mi. N on U.S. 59-71, 9 mi. E on SR 32); Ozark Lake (from Fort Smith, 35 mi. E on I-40)

CALIFORNIA

BLACK BUTTE LAKE

Situated in the rich, almond- and olive-growing region of the Sacramento River Valley in north central California, Black Butte Lake lies in a scenic area of mountains and forests. Twenty-five miles of hilly, tree-studded shoreline encircle nearly 3,000 acres of sparkling blue water. Impressive Black Butte, a part of the state's Coast Range, rises 7,450 feet within Mendocino

*Unless designated as follows, each project has restrooms, drinking water, and developed campsites. † = no restrooms; ‡ = no drinking water; § = no developed campsites.

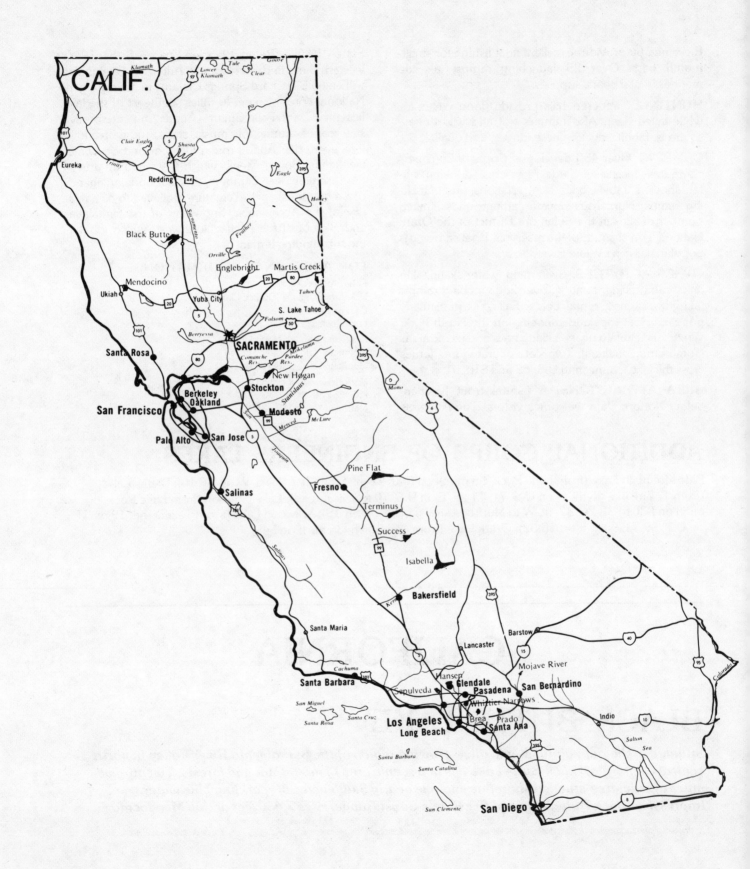

National Forest just west of the reservoir. Driftwood is abundant at Black Butte Lake, and visitors are allowed to collect as much as they want. It's available all year long, but April through June is the best time to collect it. Besides offering a wide range of water-oriented recreation at facilities which are seldom crowded, the reservoir is in an excellent centralized location from which to visit the many and varied attractions of northern California.

HOW TO GET THERE: From Orland, head 10 mi. west via Newville Rd.

FISHING: Anglers here might tie into a largemouth bass, catfish, crappie, bluegill, or striped bass. Fishing is permitted all year, but is best during early spring and summer. Three launching ramps; lakeside marina.

HUNTING: The Sacramento Valley comprises the most important wintering area for waterfowl along the Pacific Flyway. In addition to the gunning permitted on designated reservoir land, hunting is also allowed at four national wildlife refuges just south of nearby Willows—Sacramento, Colusa, Delevan, and Sutter. Though limited in size, these refuges often hold hundreds of thousands of ducks and geese from September to February. Pheasant are also hunted at Colusa. Other game species hunters will find in the area are deer, dove, rabbit, and squirrel. The Columbia black-tailed deer is a favorite quarry in the Mendocino National Forest.

CAMPING: Over 60 Corps of Engineers campsites, both primitive and developed. Facilities include restrooms, drinking water, tables, grills, and launching ramps. Additional sites in Mendocino National Forest. Private campgrounds in area include a KOA at Orland with full hookups.

OTHER ACTIVITIES: Swimming, water skiing, boating; three launching ramps, lakeside marina. Black Butte is especially popular with sailing enthusiasts. Picturesque picnic areas amidst trees and wild flowers. Several scenic drives in area. Bird-watchers might observe waterfowl, eagles, herons, egrets, hawks and, on occasion, peregrine falcons. Wildlife species which live here and are sometimes seen include fox, bobcats, beaver, deer, badger, and ground squirrels. Saddle and pack trips, as well as hiking, in the Mendocino National Forest.

AREA ATTRACTIONS: The Sacramento and Colusa national wildlife refuges near Willows offer self-guided auto tours; Colusa has an interpretive trail as well. South of Red Bluff on the Sacramento River is a public viewing plaza where visitors may watch migrating salmon and other species flip up a fish ladder. The Sacramento Valley Museum at Williams transports visitors to nineteenth century California. Table Mountain near Oroville is a volcanic lava cap which features abandoned mines, a semi-ghost town, and a museum at its peak.

FOR ADDITIONAL INFORMATION:
Public Affairs Officer
U.S. Army Engineer District, Sacramento
650 Capitol Mall
Sacramento, CA 95814

Forest Supervisor
Mendocino National Forest
420 E. Laurel St.
Willows, CA 95988

LAKE MENDOCINO

Lying along the Russian River Valley in northwestern California, Lake Mendocino is embraced by 15 miles of forested shoreline. The hills which surround the reservoir gradually rise to meet the mountains of the state's Coast Range to the east. Those towering peaks visible in the distance reach a height of 8,600 feet within Mendocino National Forest. With California's famed wine-producing region to the south, redwood forest to the north, and the ruggedly beautiful Pacific coast not far to the west, Lake Mendocino offers a variety of recreation opportunities that should please most anyone.

HOW TO GET THERE: From Ukiah, head 6 mi. north on U.S. 101, then turn east for 3 mi. on SR 20.

FISHING: A variety of warmwater fish, including striped bass, largemouth bass, crappie and channel catfish provide thrills for anglers. Trout are sometimes taken in the inlet area of the lake. Two launching ramps; lakeside marina.

HUNTING: None on reservoir lands. Columbia black-tailed deer, as well as rabbit, squirrel, quail, and migratory waterfowl, are hunted in nearby Mendocino National Forest.

CAMPING: Over 450 Corps of Engineers campsites, both primitive and developed. Facilities include restrooms, showers, drinking water, tables, grills, and launching ramps. Many private campgrounds in area; wooded sites in Mendocino National Forest.

OTHER ACTIVITIES: Swimming, water skiing, pleasure boating (including sailing); lakeside marina. Picnic areas and hiking trails along shore. Saddle and pack trips in Mendocino National Forest. Many scenic drives. Canoeing along the Russian River through groves of redwoods.

AREA ATTRACTIONS: California's most unusual train, the historic Super Skunk Steam Engine Railroad, runs from Fort Bragg to Willits and back again; it crosses 33 bridges and trestles, passes through two tunnels, and climbs from an elevation of 80 feet to 1,365 feet along a 40-mile route. The Parducci Wine Cellars north of Ukiah offer daily tours. Fifteen miles northwest of the same town, Orr's Hot Springs resort offers hot-water baths and a swimming pool. Saint Helena, in the heart of the vineyard country, has several wineries which offer tours, as well as the Hurd Beeswax Candle Factory, which demonstrates candle-making, and the Silverado Museum, which exhibits Robert Louis Stevenson memorabilia. Near Mendocino on the Pacific Coast are Van Damme and Russian Gulch state parks, where visitors may see pygmy forests of 60-year-old trees. The trees have attained a height of only a few feet because of the impoverished soil here; known as podzol, it's found in only a half-dozen places in the world.

FOR ADDITIONAL INFORMATION:

Public Affairs Officer
U.S. Army Engineer District, San Francisco
100 McAllister St.
San Francisco, CA 94102

Mendocino County Chamber of Commerce
P.O. Box 244
Ukiah, CA 95482

Forest Supervisor
Mendocino National Forest
Willows, CA 95988

NEW HOGAN LAKE

In the gently rolling foothills of northeastern California's Sierra Nevada range lies New Hogan Lake. The surrounding region, blessed with evergreen forests, sparkling streams and lakes, and magnificent vistas, is one of the most popular areas in the state for year-round recreation. Fed by the North and South forks of the Calaveras River, the reservoir is embraced by 24 miles of wooded shoreline. The most famous character ever to emerge from Calaveras County, in which New Hogan Lake is located, was a frog. Mark Twain created him in "The Celebrated Jumping Frog of Calaveras County," and every May the lucky owners of modern-day jumping frogs gather in the town of Angels Camp to vie in a competition which determines a successor to the original. Author Bret Harte also used this picturesque country as the locale for many of his writings. The area is noted, too, for one of California's most famous products—gold. State Highway 49, which runs north and south near the reservoir, passes through most of the old mining towns in the heart of mother lode country. Recreational facilities at New Hogan are maintained and administered by the Corps of Engineers.

HOW TO GET THERE: From Stockton, head 35 mi. east on SR 26 and turn south on dam road.

FISHING: Rainbow and brown trout, black and white crappie, largemouth and smallmouth bass, bluegill and channel catfish await anglers here. Bear and Haupt creeks are good areas for crappie and bass fishing. The Calaveras River is closed to fishing in the winter. Two launching ramps on lake; lakeside marina, rental boats, supplies. Additional fishing opportunities in Stanislaus National Forest not far to the east, where sportsmen will find over 715 miles of mountain streams.

HUNTING: Hunting on designated reservoir land is limited to resident and migratory game birds. Lake lies on the Pacific Flyway. Big game hunting is available in nearby Stanislaus National Forest.

CAMPING: Over 150 Corps of Engineers campsites, both primitive and developed. Facilities include restrooms, drinking water, showers, tables, grills, and launching ramps. Additional campsites in Stanislaus National Forest to the east. Many private campgrounds in area with full hookups.

OTHER ACTIVITIES: Swimming, boating, and water skiing; two launching ramps, lakeside marina, rental boats. Picnic areas in lovely settings. Over 150 species of birds have been observed on reservoir lands. The bald eagle is a winter visitor, and the sandhill cranes pause here during fall and late winter migrations. Bird-watchers will find acorn woodpeckers all year long. Scenic drives aplenty, as well as saddle and pack trips, hiking and backpacking in Stanislaus National Forest. Winter sports are extremely popular in the Sierra Nevadas.

AREA ATTRACTIONS: Some of the state's finest specimens of the gigantic Sierra redwoods may be seen at Calaveras Big Trees State Park near Murphys; the park also offers winter snowshoe hikes and cross-country ski tours. The interesting limestone formations of Mercer Caverns are near the same town. Another treat for spelunkers is near Vallecito, where strange shapes and figures adorn the walls of Moaning Caves. The most authentic and best restored of all the gold rush towns in the area is Columbia, maintained as a state historical park; visitors may pan for gold and take a stagecoach ride. Many of the gold rush towns along SR 49 feature museums. Lodi, famed as a grape and wine center, has nearly 20 wineries in the area, and most welcome visitors for tours and tasting.

FOR ADDITIONAL INFORMATION:

Public Affairs Officer
U.S. Army Engineer District, Sacramento
650 Capitol Mall
Sacramento, CA 95814

Forest Supervisor
Stanislaus National Forest
175 S. Fairview Lane
Sonora, CA 95370

Sonora Pass Vacationland, Inc.
Route 2, Box 53
Sonora, CA 95370

PINE FLAT LAKE

Located in the scenic Sierra Nevada foothills where high, steep ridges form a natural gateway to the stunning wilderness of upper Kings Canyon, Pine Flat Lake twists and turns for 20 miles along the Kings River Valley. Over half of the reservoir lies within the boundaries of the Sierra and Sequoia national forests in central California. Some of the most beautiful mountain scenery in the country is in these national forests, but much of it is accessible only by foot or horseback. Several recreation areas have been jointly developed on Pine Flat Lake by the Corps of Engineers, Fresno County, and the U.S. Forest Service.

HOW TO GET THERE: From Fresno, take SR 180 for 18 mi. to Centerville; turn northeast on lake road for 14 mi.

FISHING: Serene water in quiet coves means good fishing for smallmouth bass, white catfish, crappie, and rainbow trout. Nearly half a dozen launching ramps; lakeside marina.

HUNTING: Fairly rugged land provides winter range for California mule deer. Quail, mourning doves, bandtailed pigeons, and gray squirrels also plentiful around reservoir.

CAMPING: Over 90 Corps of Engineers campsites, both primitive and developed. Facilities include restrooms, drinking water, tables, grills, and launching ramps. Additional lakeshore campsites provided by U.S. Forest Service and Fresno County. Numerous private campsites close by.

OTHER ACTIVITIES: Swimming, water skiing, boating; lakeside marina, nearly half a dozen launching ramps. Picnicking among trees and wild flowers. Wilderness hiking and riding trails, pack and saddle trips, mountain climbing, full range of winter sports in national forests. Scenic drives on forest roads.

AREA ATTRACTIONS: Natural wonders abound in Sequoia and Sierra national forests. The Sun Maid Raisin plant in Kingsburg and several area wineries welcome visitors. In Fresno the Forestiere Underground Gardens are patterned after the catacombs of ancient Rome; trees, shrubs, and flowers grow underground. Also open to the public at Fresno are the Kearney Mansion, 1880s home of one of the county's wealthiest landowners, and the Fort Miller Blockhouse Museum in Roeding Park.

FOR ADDITIONAL INFORMATION:

Public Affairs Officer
U.S. Army Engineer District, Sacramento
650 Capitol Mall
Sacramento, CA 95814

U.S. Forest Service
California Region
630 Sansome St.
San Francisco, CA 94111

WHITTIER NARROWS LAKE

Just east of downtown Los Angeles, where the San Gabriel River and the Rio Hondo flow close together, is the Whittier Narrows Recreation Area. Normally, the reservoir here is nearly dry in order to control incoming flood flows, but the County of Los Angeles has developed two interconnected fishing lakes as part of its plan to build a major regional park here. Though many facilities have already been completed, more are planned for the future.

HOW TO GET THERE: The project is on Pomona Freeway (SR 60), west of I-605.

FISHING: Bass, bluegill, and crappie await fishermen here. Anglers may try their luck in the fishing lakes, the San Gabriel River, or Rio Hondo. Bank fishing only.

HUNTING: No hunting allowed here because of metropolitan environs.

CAMPING: None at present on project lands. Private campgrounds in area.

OTHER ACTIVITIES: Already completed are picnic areas, model car and airplane areas, four baseball diamonds, a soccer field, a trap- and skeet-shooting area, a field archery range, a rifle and pistol range, two golf courses (one 18-hole, one 9-hole), riding stables, and a children's playground. One of the most outstanding features here is the 127-acre Whittier Narrows Nature Center; the area includes a nature museum, 5 miles of trails, and a small lake. This wildlife center serves as a sanctuary for numerous species of birds, reptiles, and small mammals; and the public is invited to hear informal lectures about our environment and ecology in general. Additional features on the drawing board are a group camping area, a swimming pool complex, and tennis and handball courts.

AREA ATTRACTIONS: Nearby sights include the Mission San Gabriel, famous for its bells and museum of religious and historical treasures of early California. The Huntington Library, Art Gallery, and Oriental Gardens in San Marino are all filled with rare treasures that shouldn't be missed. Forest Lawn Memorial Park in Glendale is a cemetery that is a scenic attraction; tourists flock here to see its marble statuary and stained-glass windows (no slacks or shorts permitted, however). Football buffs may wish to tour Pasadena's Rose Bowl. Disneyland near Anaheim and Knott's Berry Farm and Ghost Town near Buena Park are not far away.

FOR ADDITIONAL INFORMATION:
Public Affairs Officer
U.S. Army Engineer District, Los Angeles
P.O. Box 2711
Los Angeles, CA 90053
Southern California Visitors Council
705 W. Seventh St.
Los Angeles, CA 90017

ADDITIONAL CORPS OF ENGINEERS LAKES*

Brea Dam§ (from Riverside Freeway, N on SR 72); Englebright Lake§ (from Marysville, 20 mi. E on SR 20); Hansen Dam§ (E of San Fernando on Foothill Blvd. SR 118); Isabella Lake (from Bakersfield, 50 mi. E on SR 178); Martis Creek Lake (from Truckee, 5 mi. E on SR 267); Mojave River Dam (from I-15, 8 mi. SE of Hesperia); Prado Dam (from Pomona, SE on SR 71); Sepulveda Dam§ (NW of U.S. 101 and I-405); Success Lake (from Portersville, 5 mi. E on SR 190); Terminus Dam (from Visalia, 20 mi. E on SR 198)

*Unless designated as follows, each project has restrooms, drinking water, and developed campsites. † = no restrooms; ‡ = no drinking water; § = no developed campsites.

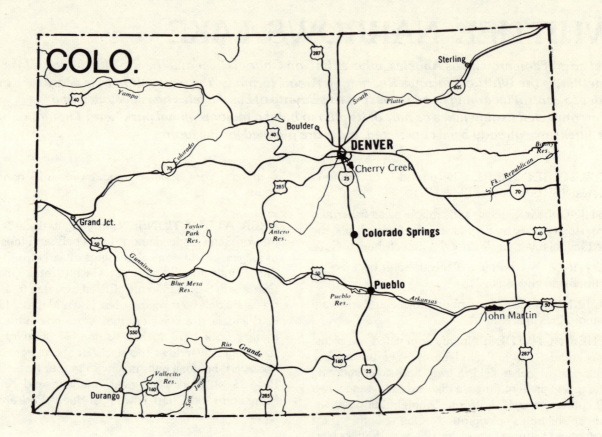

COLORADO

CHERRY CREEK RESERVOIR

Perched between the Great Plains to the east and the Rocky Mountains to the west, Cherry Creek Reservoir lies a few miles southeast of the mile-high city of Denver. It was on Cherry Creek in 1858 that the cry of "Gold!" first rang out in Colorado, heralding a gold rush that had reached stampede proportions by the following year. Today, the creek has been dammed, and its 852-acre reservoir offers a variety of seasonal activities in keeping with the state's reputation as a year-round playground. Facilities are managed by the state of Colorado.

HOW TO GET THERE: Bounded by U.S. 225 on the north and SR 83 on the east, Cherry Creek Reservoir lies on the southeast fringe of Denver.

FISHING: Fish population includes walleye, rainbow trout, northern pike, channel catfish, black bullheads, crappie, yellow perch, large and smallmouth bass, sunfish, and tench. Lakeside marina, rental boats, supplies; three launching ramps. Ice fishing in winter.

HUNTING: Primarily small game, such as rabbit, quail, and dove. There's a wildlife management area on reservoir lands. Elk, deer, and mountain sheep are found in the Pike and Arapaho national forests just west of the lake.

CAMPING: Nearly 175 developed campsites in Cherry Creek State Recreation Area. Facilities include restrooms, showers, drinking water, tables, grills, electrical hookups, dump stations and launching ramps. Many private campgrounds in vicinity.

OTHER ACTIVITIES: Swimming at sand beach, water skiing, boating; marina at lakeside, rental boats. Also picnicking, hiking, horseback riding, and bicycling. Facilities include archery, skeet and rifle ranges, and a model airplane area. A municipal golf course is near reservoir lands. The wildlife management area is a good place for wildlife observation and nature study. Winter sports of all types nearby, with ice skating on the reservoir itself.

AREA ATTRACTIONS: At Chatfield State Recreation Area southwest of Denver is a small rookery of great blue herons which has inhabited these parts for over 60 years. Near Golden, the highest auto road in the United States leads to the crest of Mount Evans, 14,260 feet high. There are many ghost towns to explore around Denver. The tomb of Buffalo Bill Cody rests atop Lookout Mountain west of Denver; a

frontiersman's museum is nearby. Tour the U.S. Mint at Denver, and while in town take a look at other attractions. Concerts by the Denver Symphony are featured in a stunning outdoor theater in Red Rocks Park west of Denver. At Idaho Springs is the Edgar Mine, used as an underground classroom for the Colorado School of Mines; visitors may take a conducted tour or search for a bit of the gold that the school dumps "over the hill" after the school is finished with it. The entire region abounds with scenic drives and beautiful mountain parks.

FOR ADDITIONAL INFORMATION:

Public Affairs Officer
U.S. Army Engineer District, Omaha
7410 U.S. Post Office and Court House
215 N. 17th St.
Omaha, NE 68102

Park Manager
Cherry Creek State Recreation Area
4201 S. Parker Rd.
Denver, CO 80232

ADDITIONAL CORPS OF ENGINEERS LAKES

John Martin Reservoir* (from Las Animas, 15 mi. E on U.S. 50 to Hasty, 2 mi. S on county road)

GEORGIA

ALLATOONA LAKE

Allatoona Lake, one of the Corps of Engineers' five most popular recreation areas, snakes its way along the Etowah River in Georgia's scenic northwestern corner. Around its 270-mile shoreline, the rolling Piedmont Plateau gradually rises on its northward journey to a meeting with the majestic Blue Ridge Mountains and the Chattahoochee National Forest. Now just off U.S. 41, the 12,000-acre lake will be crossed by I-75 when that major north-south artery is completed. Atlanta lies just an hour's drive to the southeast.

HOW TO GET THERE: From Atlanta, head north for 45 mi. on U.S. 41; turn east on SR 20 for 2 mi.; then south on SR 294 to the dam.

FISHING: Primarily warmwater species, with largemouth bass, white bass, and crappie the main attrac-

tions. Many trout streams located in the Chattahoochee National Forest. Lakeside marina, rental boats, supplies; nearly 30 launching ramps.

HUNTING: Deer, wild turkey, quail, rabbit, and squirrel abound around the reservoir and in the Chat-

*This project has restrooms, drinking water, and developed campsites.

tahoochee National Forest; hunting permitted both places.

CAMPING: Nearly 400 Corps of Engineers campsites, both primitive and developed. Facilities include restrooms, drinking water, showers, tables, grills, dump stations, and launching ramps. Red Top Mountain State Park on the lakeshore offers a large number of campsites, many with water and electrical hookups, as well as a group camping area and rental cottages. Another state park at the water's edge, George W. Carver, has a limited number of campsites.

OTHER ACTIVITIES: Swimming water skiiing, boating; nearly 30 launching ramps, lakeside marina, rental boats. Canoeing on nearby mountain streams, as well as the Etowah River. Scenic picnic areas around lakeshore. Several foot trails on reservoir lands and in state parks. Lots of backpacking opportunities in Chattahoochee National Forest.

AREA ATTRACTIONS: Some of the state's most scenic mountain drives begin at Chatsworth. Atlanta features a variety of adventures for visitors. Among them are Cyclorama, a huge three-dimensional painting in Grant Park which depicts the Battle of Atlanta during the Civil War; Underground Atlanta, a mini-community abandoned early in the century when the city's streets were raised one level to solve transportation problems of the time; and Stone Mountain Park, where likenesses of Jefferson Davis, Stonewall Jackson and Robert E. Lee on horseback have been sculpted on the face of a mountain. There's an exhibit depicting the history of the Etowah Valley in the Allatoona Reservoir manager's office, and the Etowah Indian Mounds Archeological Area is near Cartersville. Kennesaw Mountain National Battlefield Park near Kennesaw brings more Civil War history to life.

FOR ADDITIONAL INFORMATION:

Public Affairs Officer
U.S. Army Engineer District, Mobile
P.O. Box 2288
Mobile, AL 36628

Reservoir Manager
Allatoona Lake
P.O. Box 487
Cartersville, GA 30120

Forest Supervisor
National Forests in Georgia
Box 1437
Gainesville, GA 30501

JIM WOODRUFF LOCK AND DAM

The shoreline at Lake Seminole, formed by the Jim Woodruff Lock and Dam, follows an irregular 250-mile path through a rural setting of haunting beauty. Here the visitor can explore rugged ravines and hills, cypress ponds, swamps, limesinks, and lush forests of mixed pines and hardwoods. While most of its 38,000 acres of water lies in the state of Georgia, the reservoir spills over into Florida and Alabama. Over 250 islands dot its surface, and in places the water is so shallow it can be waded a mile from shore. The Jim Woodruff Lock and Dam stretch across the Apalachicola River just below its junction with the Chattahoochee and Flint rivers, which form the major branches of Lake Seminole. Each year about 3 million people visit this popular Corps of Engineers project to enjoy a variety of recreational activities.

HOW TO GET THERE: From Tallahassee, FL, head 42 mi. west on U.S. 90 to Chattahoochee; turn north for 1 mi. to dam.

FISHING: Called by many the largemouth bass capital of Georgia. The shallow waters, full of standing dead timber and cypress trees, commonly produce lunkers of 15 pounds and over. Other species include bluegill, crappie, red ear sunfish, catfish, and chain pickerel. Good fishing all year, but February, March, and October are the top bass months. Almost 35 launching ramps; lakeside marina, rental boats, supplies.

HUNTING: In designated areas around the reservoir, including the Apalachee Game Management Area along the lakeshore in Florida; also in Apalachicola National Forest and the adjoining Aucilla Wildlife Management Area in the Sunshine State. Deer and quail hunting are top-notch. Other species available include wild turkey, dove, rabbit, squirrel, and waterfowl.

CAMPING: Nearly 150 Corps of Engineers campsites, both primitive and developed. Facilities include restrooms, drinking water, showers, picnic tables, grills, and launching ramps. Seminole State Park in Georgia and Three Rivers State Park in Florida both have numerous modern campsites. Private campgrounds in area.

OTHER ACTIVITIES: Lake swimming, water skiing, boating; lakeside marina, rental boats, nearly 35 launching ramps. Canoeing on Florida's Chipola River and on streams within Apalachicola National Forest. Hiking, picnicking, bird-watching and nature study are all popular. Outstanding variety of plant, bird, and animal species has received national recognition. The national forest offers backpacking, horseback riding, and scenic drives; along the way visitors may see beehives which are part of a local honey production operation and groups of "worm grunters" (individuals engaged in gathering worms to be sold commercially by driving notched sticks into the ground and rubbing another stick rhythmically across the notches, thus setting up reverberations underground which drive worms to the surface).

AREA ATTRACTIONS: (Florida) Near Marianna, you can visit Florida Caverns State Park and tour the Sunshine State's largest cave. Maclay Gardens are near Tallahassee. Wakulla Springs offers glass-bottomed boat rides in one of the largest spring basins in the world. (Georgia) Thomasville has some of the most famous roses in the country and a long season in which to view them at the Rose Test Gardens. The same town is also noted for its magnificent plantations and the sprawling "Big Oak." Howard's Grist Mill near Jakin has been water-grinding meal and grits for over 125 years.

FOR ADDITIONAL INFORMATION:

Public Affairs Officer
U.S. Army Engineer District, Mobile
P.O. Box 2288
Mobile, AL 36628

Reservoir Manager
Lake Seminole
P.O. Box 96
Chattahoochee, FL 32324

Historic Chattahoochee Commission
P.O. Box 33
Eufaula, AL 36027

Forest Supervisor, Apalachicola
National Forests in Florida
Box 1050
Tallahassee, FL 32302

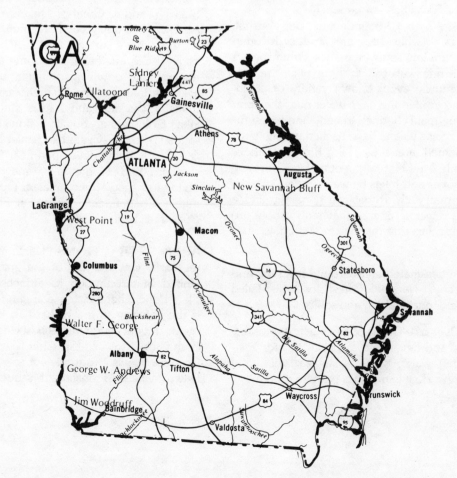

LAKE SIDNEY LANIER

A lovely lake of many facets, Lake Sidney Lanier lies in a land steeped in history, legend, and folklore. It was created in 1957 by Buford Dam on the Chattahoochee River. Within a few short years, it had become the most popular Corps of Engineers' reservoir in the entire country, a status this north Georgia lake has consistently maintained to this day. Visitation now totals nearly 15 million people annually. The reservoir's 540-mile shoreline varies from deep pine woodlands to rocky outcroppings to sandy beaches. A conservation pool of 38,000 acres, its surface dotted with many islands, backs up the broad valley for 40 miles and spreads its waters into countless coves along the way. In the distance, to the north and the west, can be seen the peaks of the Blue Ridge Mountains. Also to the north is the rugged Chattahoochee National Forest, whose mountainous watersheds give rise to the Chattahoochee River. The Cherokee Indians loved this land long before Georgia poet Sidney Lanier immortalized it in Song of the Chattahoochee. *Though portions of Lake Lanier retain a semiwilderness atmosphere even today, other parts offer every modern convenience; and Atlanta is just an hour's drive to the southwest.*

HOW TO GET THERE: From Atlanta, take I-85 north for nearly 30 mi. to SR 365, then turn northwest. All west exits from Exit 3 on lead to the project. SR 365 at present ends at Gainesville on the lakeshore.

FISHING: Lake Lanier fishing has over the years developed into some of the best in the Southeast. Largemouth bass catches are generally larger in size than in number; 7- to 10-pounders are not unusual. Large numbers of white bass are produced during spring and fall runs into some tributaries. Other species in the lake include walleye, northern pike, crappie, catfish, and virtually every known variety of perch. There are some trout in the deep water near the dam, as well as in the main channels in early and late summer. The reservoir's waters have backed up into such intriguingly named areas as Bald Ridge, Wahoo, Flowery Branch, Mud Flat, and Young Deer, forming hundreds of coves and inlets between fingers of land, and the fisherman with a boat can generally find a private spot to wet a line somewhere. About 50 launching ramps; lakeside marinas, rental boats, supplies.

HUNTING: In designated areas around reservoir and in Chattahoochee National Forest for white-tailed deer, wild turkey, quail, rabbit, and squirrel.

CAMPING: Over 520 Corps of Engineers campsites, both primitive and developed. Facilities include restrooms, drinking water, showers, tables, grills, and launching ramps. Boat camping is permitted on many islands. Private campgrounds on the lakeshore provide water and electrical hookups.

OTHER ACTIVITIES: Lake swimming at several sand beaches, three municipal swimming pools in Gainesville, water skiing, boating; lakeside marinas, rental boats, about 50 launching ramps. Lake Lanier claims two titles: "Houseboat Capital of the United States" and "Greatest Inland Sailing Fleet in America." At any given time, there are plenty of each on the water's surface. On the Chattahoochee River above Lake Lanier, there's some fine whitewater canoeing. And down near Atlanta, visitors and natives alike are taking to the Chattahoochee in rubber inner tubes. Picnicking and nature study all around the lake. Golf, tennis, bowling, horseback riding in the vicinity of Gainesville. Lake Lanier Islands Resort Area is currently being developed near Buford Dam at the southern end of the lake, and facilities are being opened as they are completed. Opportunities for hiking in Chattahoochee National Forest; the Appalachian Trail begins here.

AREA ATTRACTIONS: Dahlonega, north of the reservoir, was the site of the first gold rush in the United States; among its attractions are Blackburn State Park, where visitors can pan for gold. Cleveland is the gateway to the Richard B. Russell Parkway, a scenic drive through the highest elevations in the state. A memorial tower atop Chenocetah Mountain near Cornelia allows one to view four states on clear days. Dawsonville's Moonshine Museum displays an authentic still. The town of Helen features a simulated Bavarian Alpine village in a peaceful mountain valley. In Atlanta's Grant Park is Cyclorama, a huge three-dimensional painting depicting the Battle of Atlanta during the Civil War. Underground Atlanta is the restoration of a minicommunity abandoned early in the century when Atlanta's streets were raised one level to solve transportation problems of the time. East of Atlanta is Stone Mountain Park, where likenesses of Jefferson Davis, Stonewall Jackson, and Robert E. Lee have been sculpted on the face of a mountain.

FOR ADDITIONAL INFORMATION:
Public Affairs Officer
U.S. Army Engineer District, Mobile
P.O. Box 2288
Mobile, AL 36628

Resource Manager
Lake Sidney Lanier
P.O. Box 567
Buford, GA 30518

Forest Supervisor
National Forests in Georgia
Box 1437
Gainesville, GA 30501

WEST POINT LAKE

In west central Georgia, where the Chattahoochee River meets the Alabama border after winding its way across the Peach State from its Blue Ridge Mountain origin, lies beautiful West Point Lake. Lush green forests, placid meadows, and gently rising hills surround 26,000 acres of deep blue water. In the upper reaches of the reservoir, large areas of trees have been left standing to improve fishing and waterfowl hunting. One of the Corps of Engineers' newest and most innovative projects, West Point Lake has been designed as a recreation showcase, with conservation of wildlife and natural beauty of major concern. Plans for the future include a Corps-operated conservation-education center for use in outdoor education programs. The handicapped will have two day-use areas developed especially for them, and they will have access to and provisions for use of all facilities. A relocated Chattahoochee River bridge, over SR 219, was adjudged prize bridge in nationwide competition among spans of the same type. In addition to those instituted by the Corps of Engineers, recreational facilities along the 525-mile shoreline will be provided by three state parks, one county park, four municipal and county access areas, and a $5 million development by the Burnt Village Park Authority, which is sponsored by the state of Alabama. At present, these facilities range from the planning to finished stages, with overall completion programed for 1977.

HOW TO GET THERE: From La Grange, GA, take U.S. 29 west and south for approximately 15 mi. to the dam.

FISHING: Although still too young to be a good producer, West Point Lake is predicted to become a largemouth bass paradise. Anglers may also expect a good population of channel catfish, white bass, crappie, bream, and sunfish. Preteen children will have their own 2.5-acre fishing pond near a downstream access area. Over 30 launching ramps; marinas on lake. Several fishing piers planned.

HUNTING: The state of Georgia operates a wildlife

management area of nearly 7,000 acres on land adjoining the reservoir's upper reaches. Hunters will find deer, wild turkey, small game, and game birds. Waterfowl hunting is expected to be improved by large areas of trees left standing when reservoir waters rose and by the construction of several duck ponds. A small-bore rifle and pistol range is planned here.

CAMPING: Initially the Corps of Engineers is developing 26 recreational areas which will contain 600 camping sites. Many of these have been completed and are open to the public. Facilities include restrooms, showers, drinking water, tables, grills, dump stations, some water and electrical hookups, and launching ramps. Several group camping areas planned. A modern campground with well over 100 campsites is available at Burnt Village Park, Alabama.

OTHER ACTIVITIES: Lake swimming, water skiing, boating; lakeside marina, boat rentals, supplies, over 30 launching ramps. Swimming pool at Burnt Village Park in Alabama. Picnic areas, hiking trails, two amphitheaters, playgrounds, baseball diamonds, tennis-basketball courts, and golf course.

AREA ATTRACTIONS: (Georgia) Callaway Gardens near Pine Mountain are outstanding; in addition to the many natural attractions, the gardens are the site of annual summer shows presented by the Florida State University Flying High Circus. At Warm Springs is the Little White House of President Franklin D. Roosevelt and a national fish hatchery. The Fort Benning Infantry Museum and a Confederate Naval Museum are at Columbus. Near Lumpkin is Westville, a restored 1850 village where today's craftsmen demonstrate the skills of yesteryear. (Alabama). Horseshoe Bend National Military Park near Dadeville commemorates a major victory over the Creek Indian nation in 1814; the triumphant Tennessee Militia brought national fame to itself and its leader, Andrew Jackson.

FOR ADDITIONAL INFORMATION:

Public Affairs Officer
U.S. Army Engineer District, Mobile
P.O. Box 2288
Mobile, AL 36628

Resource Manager
West Point Lake
P.O. Box 574
West Point, GA 31833

Burnt Village Park Campground
P.O. Box 205
Lanett, AL 36863

Historic Chattahoochee Commission
P.O. Box 33
Eufaula, AL 36027

ADDITIONAL CORPS OF ENGINEERS LAKES*

George W. Andrews Lake (from Albany, 60 mi. W on SR 62 to Hilton, signs to dam); Walter F. George Lake (from Albany, 24 mi. W on SR 62 to Leary, 36 mi. W on SR 37 to Fort Gaines, 2 mi. N on SR 39 to dam)

IDAHO

ALBENI FALLS RESERVOIR (LAKE PEND OREILLE)

The largest lake in Idaho, Lake Pend Oreille has a 226-mile shoreline draped with forests and touched by four mountain ranges. An impoundment of the Pend Oreille River in the Gem State's

*These projects all have restrooms, drinking water, and developed campsites.

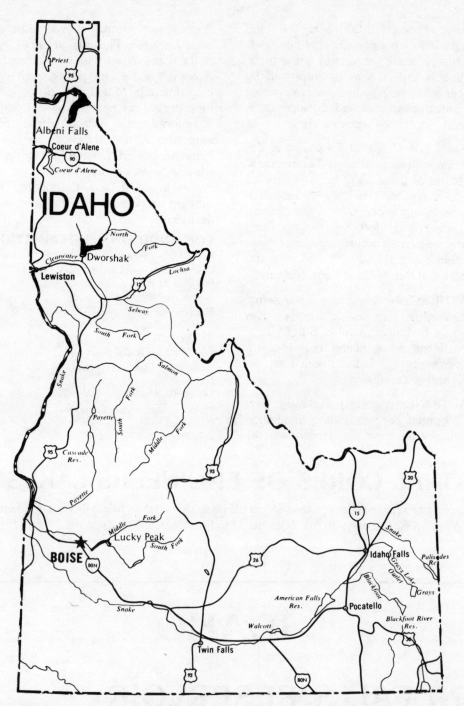

panhandle, it is noted for its scenic fiords and clear, deep waters. This was the site chosen for the first American-based Boy Scout World Jamboree. At the water's edge are two state parks and the Kaniksu National Forest, all of which offer recreational complexes to complement Corps of Engineers facilities.

HOW TO GET THERE: From Spokane, WA, head northeast for 50 mi. on U.S. 2.

FISHING: This lake's greatest claim to fame, by far, is the fish it produces. The world record kamloops rainbow trout and the world record Dolly Varden both came from its waters, and there are other giants still lurking there. Another popular species is the land-locked sockeye salmon, and there are cutthroat and brown trout, bass, crappie, perch, sunfish, and white-fish as well. Nearly half a dozen launching ramps; rental boats and supplies. Ice fishing in winter.

HUNTING: Migratory birds, primarily mallards and Canada geese, are plentiful for waterfowl gunners. Deer, elk, moose, antelope, bighorn sheep, and mountain goat roam the forests and mountains. Hunt-ing along shoreline of reservoir and banks of Pend Oreille River in designated areas; also in several nearby wildlife management areas and Coeur d'Alene and Kaniksu national forests.

CAMPING: Nearly 150 Corps of Engineers developed campsites; facilities include restrooms, drinking water, tables, grills, firewood, and launching ramps. Additional modern sites at Sunnyside and Far-ragut state parks and in the Samowen Recreation Area operated by the U.S. Forest Service. Several private campgrounds nearby.

OTHER ACTIVITIES: Lake swimming, a children's wading and swimming pool, water skiing, boating and sailing; rental boats, nearly half a dozen launching ramps. Picnicking, hiking and bike trails, rockhound-ing, wildlife observation, nature study, scenic drives, horseback riding, and mountain climbing. Canoeing on Priest River. Full range of winter sports. The public beach at Sandpoint has a playground and tennis court.

AREA ATTRACTIONS: Kaniksu National Forest fea-tures the rugged backcountry of the Selkirk Mountain Range, the Roosevelt Ancient Grove of 800-year-old cedars, and unusual Chimney Rock. In nearby Coeur d'Alene National Forest are several large sawmills, a mission of 1846 vintage, and a rich mining district which produces zinc, lead, and silver. Coeur d'Alene Indian Reservation partially encompasses Coeur d'Alene Lake, rated as one of the world's five most beautiful; visitors may see this lake more intimately via excursion boat, mail boat, or seaplane. The Kootenai National Wildlife Refuge is near Bonners Ferry. Several ghost towns are eerie reminders of fabulous mineral strikes made in a bygone era. Spokane, with its many attractions, is 50 miles away.

FOR ADDITIONAL INFORMATION:

Public Affairs Officer
U.S. Army Engineer District, Seattle
4735 E. Marginal Way S.
Seattle, WA 98134

Forest Supervisor
Kaniksu National Forest
P.O. Box 490
Dover Highway
West of Sandpoint, ID 83864

Forest Supervisor
Coeur d'Alene National Forest
P.O. Box 310
Coeur d'Alene, ID 83814

DWORSHAK RESERVOIR

A 53-mile long lake which extends back into the wild, rugged timber land of north central Idaho, Dworshak Reservoir begins within the Nez Percé Indian Reservation. In every direction mountains with forests on their backs march to the horizon. Not far to the east is Clearwater National Forest. Over 5,000 acres of Corps of Engineers land in the upper reaches of the reservoir will be managed exclusively for elk habitat. Visitors are almost always impressed by the massive Dworshak Dam, the largest of its type in the Western World; it's particularly beautiful when lighted at night. Because the North Fork of the Clearwater River is so valuable as a shipping lane for the area's timber harvest, permanent log-handling facilities were built at the dam; and visitors may have the opportunity to see them at work.

HOW TO GET THERE: From Lewiston, take U.S. 12 east for 45 mi.

FISHING: Dworshak is stocked annually with rainbow and cutthroat trout and kokanee salmon.

Deer are found at many lakeside areas.

Dolly Varden trout and smallmouth bass are native fish here. Nearly half a dozen launching ramps. Fuel and fishing equipment is seasonally available at Big Eddy, a concession operated by the Nez Percé Indian Tribal Council.

HUNTING: Elk and white-tailed deer are found on reservoir land, with ruffed grouse, blue grouse, mourning doves, and cottontail rabbits adding small game action. Deer and elk browse is being developed on acreage adjacent to the lake.

CAMPING: At present, just primitive camping is available here. All 200 sites are operated by the Corps of Engineers and are accessible by boat. Each site has a picnic table, fireplace, leveled tent pad, and chemical flush toilet. Private campgrounds nearby, including a KOA with a swimming pool and full hookups at Orofino.

OTHER ACTIVITIES: Swimming, water skiing, and boating; about half a dozen launching ramps. Prime area for wildlife observation; inhabiting the surrounding countryside are black bear, bald and golden eagles, osprey, coyote, bobcat, porcupines, chipmunks, and a variety of songbirds. Very scenic picnic areas. Hiking through evergreen forests. Boaters and campers may pick an assortment of wild berries in season. Because migration of steelhead trout was blocked by construction of Dworshak Dam, the Corps of Engineers built the world's largest steelhead fish hatchery just downstream. Canoeing on the Clearwater River and its three forks. Full range of winter sports found in surrounding region. Precious and semiprecious stones in the area delight rockhounders. U.S. 12 and North Fork Road are scenic routes through Clearwater National Forest.

AREA ATTRACTIONS: The Nez Percé National Historical Park near Spalding is made up of 22 separate sites of historical significance, including the battlefield where the famous Nez Percé War began in 1877. The world's largest white pine mill in Lewiston offers guided tours. Snake River boat trips to the entrance of famous Hell's Canyon originate in Lewistown. Just north of here at Moscow is the beautiful campus of the University of Idaho, recognized as a landscaping masterpiece. Mount Idaho, southeast of Grangeville, once a flourishing mining community, is now a ghost town.

FOR ADDITIONAL INFORMATION:
Public Affairs Officer
U.S. Army Engineer District, Walla Walla
Bldg. 602, City-County Airport
Walla Walla, WA 99362

ADDITIONAL CORPS OF ENGINEERS LAKES

Lucky Peak Lake* (from Boise, 10 mi. SE on SR 21)

ILLINOIS

CARLYLE LAKE

A vast 26,000-acre reservoir in the wide-open spaces of south central Illinois, Carlyle is the Prairie State's largest man-made lake. Here visitors can drink their fill of small rural communities and the calm beauty of prairies and farmlands while enjoying some of the finest water-based recreation in the entire Midwest. For a change of pace, St. Louis sprawls along the banks of the Mississippi River 50 miles to the west. There are two state parks at the water's edge to complement Corps of Engineers recreation areas.

---

*This project has restrooms, drinking water, and developed campsites.

HOW TO GET THERE: From St. Louis, head east on U.S. 50 for approximately 50 mi.

FISHING: Most extensive stocking is comprised of largemouth and white bass. Crappie fishing is tops. Other species adding to the catch are bluegill, bullheads, channel catfish, yellow bass, sunfish, sauger, bowfin, carp, and freshwater drum. Bank fishing is particularly productive both above and below the dam, while boaters will find many coves and inlets as well as deep hole beds. A dozen launching ramps service the lake; marinas, rental boats, and supplies at lakeside.

HUNTING: Migratory waterfowl stop here as they travel the Mississippi Flyway. Hunters also seek out dove, squirrel, rabbit, quail, and deer. Hunting allowed on 16,000-acre Carlyle Lake Wildlife Management Area at the north end of the reservoir and on designated Corps of Engineers land.

CAMPING: Around 350 Corps of Engineers campsites, both primitive and developed. Facilities include restrooms, drinking water, tables, grills, some showers and electrical hookups, firewood, playgrounds, dump stations, and launching ramps. Over 300 more campsites at Eldon Hazlett and South Shore state parks, many with hookups. Many private campgrounds in surrounding area.

OTHER ACTIVITIES: Lake swimming, water skiing, boating; lakeside marina, rental boats, a dozen launching ramps. The lake has relatively shallow water and winds that make it ideal for sailing; the Hazlett State Park Marina is the center of sailing activity, with regattas and competitive events held regularly. The Kaskaskia River attracts canoeists. The lake area here has been carefully planned to ensure an abundant wildlife population and preservation of the natural environment; visitors may view both from picnic areas and hiking trails.

AREA ATTRACTIONS: Several Indian mounds and a museum may be viewed at Cahokia Mounds State Park near East St. Louis. The National Shrine of Our Lady of the Snows at Belleville attracts nearly a million people of all faiths each year; the gardens and reflection pool are impressive. Okawville was once called the "Little Hot Springs of Illinois"; mineral baths, Swedish massages, and gourmet food are still featured there. A visit to a 225-ton steam locomotive on display at Centralia will delight the small fry. They'll also enjoy the swinging foot bridge, named in honor of Korean War General Dean, just east of Carlyle. The cross outside Breese is dedicated to the victims of the 1849 cholera epidemic. St. Louis, an hour's drive away, is certainly worth exploring. Lincoln served in the legislature at what is now the Statehouse State Memorial in Vandalia; there are several places of interest to history buffs.

FOR ADDITIONAL INFORMATION:
Public Affairs Officer
U.S. Army Engineer District, St. Louis
210 N. 12th St.
St. Louis, MO 63101
Eldon Hazlett & South Shore State Parks
1351 Ridge St.
Carlyle, IL 62231

LAKE SHELBYVILLE

In the midst of some of America's richest farmland, along the fertile, tree-lined valley of Illinois' Kaskaskia River, lies 11,000-acre Lake Shelbyville. Over a century ago, a tall, thin man walked this way and cast a giant shadow which remains undiminished to this day. This is Lincoln country, and his enduring legend dominates much of the surrounding countryside. It was in the town of Shelbyville that Lincoln began the debating career which eventually led him to national prominence. Today that same town is the center of a bustling tourist industry, and over 2.5 million visitors are annually attracted to the recreational waters of Lake Shelbyville.

HOW TO GET THERE: The project's dam is at the town of Shelbyville, located southeast of Springfield on SR 16.

FISHING: Largemouth and smallmouth bass, walleye, and northern pike are stocked extensively. Other species include white bass, crappie, catfish, bull-

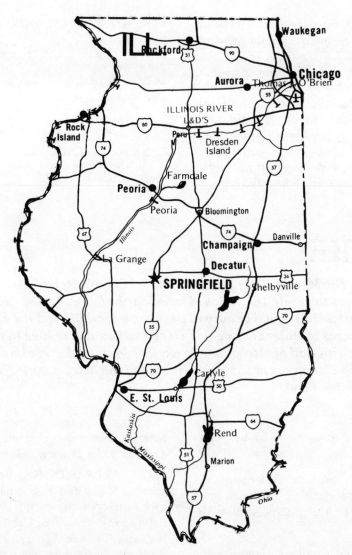

head, bluegill, sunfish, drum, carp, and bowfin. Nearly 10 launching ramps available; lakeside marina, rental boats, supplies.

HUNTING: Migratory waterfowl are important quarry here. Also dove, squirrel, quail, rabbit, pheasant, deer, and woodcock. About 6,000 acres of land and water in the upper reaches of the reservoir have been designated as wildlife management areas.

CAMPING: A mix of Corps of Engineers, state park, and privately managed facilities provide an abundance of campsites on reservoir land. Nearly 800 Corps of Engineers sites, both modern and primitive; facilities include restrooms, showers, drinking water, tables, grills, playgrounds, dump stations, some electrical hookups, and launching ramps. Two state parks, Eagle Creek and Wolf Creek, are on opposite shores

of the reservoir; there are nearly 500 state-operated campsites, some with electrical hookups. Private sites on reservoir land total nearly 120.

OTHER ACTIVITIES: Swimming, water skiing, boating of all types; marina and rental boats. Canoeing on the reservoir and on the Kaskaskia River. Numerous picnic sites and hiking trails. Indian artifacts can be found in a number of old camps along the Kaskaskia River in this region.

AREA ATTRACTIONS: Shelbyville's Washington Street offers views of beautifully preserved Civil War homes. Lincoln Log Cabin State Park near Charleston contains the cabin home of Lincoln's father and stepmother; both are buried at nearby Shiloh Cemetery. Lincoln served in the legislature at Vandalia, where the Statehouse State Memorial preserves the old state capitol building. Springfield, 60 miles away, will be of interest to history buffs. There are many interesting attractions, but this is primarily a city which honors the memory and accomplishments of our sixteenth president. Sullivan boasts a highly successful summer stock theater which regularly lures top stars from Hollywood, Broadway, and television land. An interesting mix is offered at the Rockome Gardens and Old Bagdad Town near Arcola; on the grounds are flower gardens and rockwork displays, an authentic Amish house, antique museum, and a haunted barn.

FOR ADDITIONAL INFORMATION:
Public Affairs Officer
U.S. Army Engineer District, St. Louis
210 N. 12th St.
St. Louis, MO 63101

Lake Manager
Lake Shelbyville
P.O. Box 26
Shelbyville, IL 62565

Eagle Creek and Wolf Creek State Parks
103 W. South 2nd St.
Findlay, IL 62534

REND LAKE

An impoundment of the Big Muddy River in south central Illinois, Rend Lake reaches into numerous inlets and coves to create 162 miles of forested shoreline. To the north is land rich in Lincoln lore; to the south is the rugged beauty of the Ozark foothills and the Shawnee National Forest. Nearly 19,000 acres of water surface povide recreational activities here, while the surrounding area offers a myriad of attractions to explore from a lakeside base. There are several Corps of Engineers recreation areas at the southern end of the lake; Wayne Fitzgerrell State Park borders much of the northern end.

HOW TO GET THERE: Just west of I-57 between Benton and Mount Vernon.

FISHING: Fishing variety is provided by largemouth bass, bluegill, sunfish, crappie, catfish, bullhead, and carp. Lakeside marina, rental boats, supplies; over 15 launching ramps.

HUNTING: Hunters will find waterfowl, squirrel, rabbit, deer, woodcock, quail, dove, and pheasant. Two large subimpoundment areas at the north end of Rend Lake have been established for the management of wildlife; public hunting is also permitted in other designated project lands.

CAMPING: Around 400 Corps of Engineers campsites, both primitive and developed; facilities include restrooms, drinking water, tables, grills, playground, dump stations, and launching ramps. Approximately 150 primitive sites in Fitzgerrell State Park. Excellent private campgrounds are close by.

OTHER ACTIVITIES: Swimming, water skiing, boating of all types; lakeside marina, rental boats, over 15 launching ramps. Picnic areas and hiking trails around reservoir. Naturalist programs at the Corps' Dam West and Sandusky Creek access areas.

AREA ATTRACTIONS: Shawnee National Forest consists of forested, rolling hills, lakes, and spectacular rock outcroppings which invite hikers, sightseers, and nature lovers. South of Carterville is the Crab Orchard National Wildlife Refuge, a unique area which achieves the compatibility of wildlife with industrial and recreational operations within its boundaries. Abraham Lincoln practiced law in the Old Appellate Courthouse in Mt. Vernon, one-time home of the State Supreme Court; it's noted today for its extensive legal library. On SR 148 near Scheller Lake is the site of the world's largest slope mine. The Crab Orchard and Egyptian Railroad in Marion offers sightseeing rides on old-fashioned Pullman cars.

FOR ADDITIONAL INFORMATION:
Public Affairs Officer
U.S. Army Engineer District, St. Louis
210 N. 12th St.
St. Louis, MO 63101

Wayne Fitzgerrell State Park
P.O. Box D
Benton, IL 62812

Corps of Engineers lakes often attract geese. (Illinois Dept. of Conservation photo)

ADDITIONAL CORPS OF ENGINEERS LAKES*

Farmdale Dam†, ‡, § (2 mi. SE of East Peoria); the following locks and dams on the Illinois Waterway: Thomas J. O'Brien Lock§ (in Chicago, S on I-94, E on 130th St., follow signs); Dresden Island Lock and Dam§ (from Chicago, S on I-55, W on SR 6); Peoria Lock and Dam§ (S of Peoria); La Grange Lock and Dam§ (S of La Grange)

---

*Unless designated as follows, each project has restrooms, drinking water, and developed campsites. † = no restrooms; ‡ = no drinking water; § = no developed campsites.

INDIANA

CECIL M. HARDEN LAKE (MANSFIELD RESERVOIR)

Formerly known as Mansfield Reservoir, and often referred to as Raccoon Lake by nearby residents, this lovely, picturesque reservoir was recently renamed Cecil M. Harden Lake by an act of Congress. The name honors a U.S. representative from the Hoosier State who was instrumental in gaining funds for the project. By any name, the lake remains a favorite spot for water recreation in west central Indiana. Harden Lake is located on Raccoon Creek in Parke County, which claims more covered bridges than any other county in the entire United States. This fact, along with the local maple syrup industry and forested hills which burst into flaming colors in the fall, creates an aura of transplanted New England. Recreational facilities along the 2,100-acre reservoir are administered by the state, and there are two state parks nearby.

HOW TO GET THERE: From Rockville, head east on U.S. 36 for about 10 mi.

FISHING: Anglers will find largemouth and smallmouth bass, catfish, bluegill, crappie, and walleye. Nearly half a dozen launching ramps; lakeside marina, rental boats, supplies.

HUNTING: Deer, squirrel, rabbit, quail, and pheasant are stalked by many hunters. Hunting in designated areas around reservoir and in nearby Owen-Putnam State Forest.

CAMPING: Around 200 campsites, both primitive and modern, in the Raccoon Lake State Recreation Area along the lake. Facilities include restrooms, drinking water, showers, tables, grills, firewood, dump station, and launching ramps. Additional campsites in Turkey Run and Shades state parks nearby.

OTHER ACTIVITIES: Lake swimming at sand beach, water skiing, boating of all types; lakeside marina, rental boats, nearly half a dozen launching ramps. Shaded picnic areas. Some outstanding hiking trails around reservoir and in nearby area; Turkey Run and Shades state parks feature rugged terrain and virgin timber stands. Rent a bicycle at the state recreation area and tour Parke County's backroads. Over 20 miles of bridle paths at Turkey Run State Park, where horses can be rented; the same park also has a tennis court. Scenic drives include four covered bridge routes in Parke County; bus tours are also available. Canoeing on Wabash River.

AREA ATTRACTIONS: Billie Creek Village near Rockville is a pioneer restoration featuring local arts and crafts. Rockville is also the headquarters for a Covered Bridge Festival in the fall and a Maple Syrup Festival in the spring. A water-powered grist mill near Mansfield still operates during summer months. There's a monument and memorial park honoring author Ernie Pyle near Dana, his hometown. Near Riley are preserved remnants of the Wabash and Erie Canal, once the longest canal in the country. Indianapolis is less than an hour's drive away. Among its attractions are the Indianapolis Motor Speedway and Museum, open to the public; an outstanding Children's Museum, and the homes of Hoosier poet James Whitcomb Riley and President Benjamin Harrison.

FOR ADDITIONAL INFORMATION:
Public Affairs Officer
U.S. Army Engineer District, Louisville
P.O. Box 59
Louisville, KY 40201

MONROE LAKE

Indiana's largest man-made lake, situated in the south central part of the state, features a scenic stone-bluffed shoreline partially encompassed by the wooded hills of the Hoosier National Forest. It lies in the heart of the state's most striking scenery, a region of deep valleys, sharp ridges, caves, sinkholes, and disappearing streams. An impoundment of Salt Creek, the reservoir offers 11,000 acres of surface water which provide water-based recreation for over 1.25 million visitors annually. Most of the public facilities are administered by the U.S. Forest Service or the state of Indiana. In addition, there are numerous privately owned businesses around the lake which cater to visitors. Inn of the Fourwinds is a strikingly beautiful luxury motel, owned by the Ramada Inn chain and offering every modern convenience.

HOW TO GET THERE: From Bloomington, head south on SR 37 and exit at Smithville Rd. or Harrodsburg exit. The reservoir lies east of the highway.

FISHING: Known as one of the Midwest's best fishing holes. There have been several largemouth bass catches in excess of eight pounds, and bluegill fishing is rated excellent. Smallmouth, spotted, and rock bass; channel, flathead, and blue catfish; crappie and northern pike are also stocked here. Several bays with flooded timber around the shoreline are especially good for ice fishing. Nearly a dozen launching ramps around the 137-mile shoreline; lakeside marinas, boat rentals and supplies.

HUNTING: The white-tailed deer is a favorite of sportsmen, but exciting sport is also provided by squirrels, rabbits, foxes, raccoon, quail, wild turkey, and waterfowl. Small streams around the reservoir offer the opportunity for floating jump-shoots; ducks are the quarry. The Brown County grouse are nationally famous for their elusiveness.

CAMPING: Over 225 state-owned and more than 325 Forest Service campsites, with facilities ranging from primitive to the most modern. Numerous private campgrounds near lakeshore.

OTHER ACTIVITIES: Several swimming beaches around the lake. Water skiing in designated areas. Pleasure boating of all types; houseboating, sailing, and canoeing are becoming increasingly popular. Lakeside marinas offer rental boats and supplies. Picnic areas rim the lake. Hiking trails in the surrounding countryside are numerous; visitors will see a variety of natural attractions along them, including a tract of virgin woods, some impressive waterfalls (in the spring), and interesting rock formations. Horseback riding, with stables and trails; nearby Brown County State Park has an archery range and a small nature museum. Rent a bicycle and take to the backroads.

AREA ATTRACTIONS: Indiana University at Bloomington is internationally famous for its music school, and it offers a variety of public programs throughout the year. Lilly Library on the campus has a collection of rare books and manuscripts. Nearby Brown County has an aura reminiscent of the Great Smoky Mountains; the town of Nashville is an artist's colony with some interesting shops unique to this part of the country. Visitors can view the world's largest limestone quarries north of Bedford. West of the same town, Purdue University operates an experimental farm which welcomes onlookers. Ozark Fisheries near

Martinsville produces over 40 million goldfish annually; visitors will enjoy the waterplants which grow in nearly 1,000 ponds. Spring Mill State Park near Mitchell features excursions into the past and the future; Spring Mill Village, of 1815 vintage, is one of the most successful historical restorations in the United States, while the Grissom Memorial Visitors Center offers a space capsule and movie on space exploration.

FOR ADDITIONAL INFORMATION:
Public Affairs Officer
U.S. Army Engineer District, Louisville
P.O. Box 59
Louisville, KY 40201
Forest Supervisor
Wayne-Hoosier National Forest
1615 J Street
Bedford, IN 47421

SALAMONIE LAKE

This narrow, deep reservoir in northeastern Indiana offers nearly 3,000 acres of water playground at the edge of the state's rich corn belt. The shoreline, partially blanketed with trees, partially met by broad meadows, is bordered by state recreation areas at each end of the lake. Monument Island is a year-round wildlife refuge located in the lake's largest cove. Below the dam, Salamonie River flows northwest for three miles to its junction with the lovely Wabash. Though visitors will find many outdoor activities here, facilities are rarely crowded. Two other Corps of Engineers reservoirs, Huntington and Missinewa, are within a half-hour's drive and offer additional recreation opportunities.

HOW TO GET THERE: From Wabash, head south for 4 mi. on SR 15. Turn east on SR 124 for 9 mi.; then head north on SR 105 to reservoir.

FISHING: Stocked with largemouth, smallmouth, and rock bass, channel cats, flathead catfish, bluegill, black crappie, and red ear sunfish. Over half a dozen launching ramps; lakeside marina, rental boats, supplies. Ice fishing in winter.

HUNTING: Deer, quail, pheasant, squirrel, and rabbit call this home. Migrating waterfowl provide one of the finest hunting experiences here.

CAMPING: The state of Indiana has established over 500 campsites, both modern and primitive. Facilities include restrooms, drinking water, showers, tables, grills, playground, dump station, and launching ramps.

One group camping area. Private campgrounds in nearby area.

OTHER ACTIVITIES: Lake swimming, water skiing, boating; lakeside marina, rental boats, over half a dozen launching ramps. Picnicking, hiking and riding trails, ice skating in winter.

AREA ATTRACTIONS: Near Lagro are the remains of old canal locks and Hanging Rock, an interesting formation carved by river erosion. The Dora Covered Bridge is on reservoir lands. Northwest of Bluffton is Deam Oak, a rare natural hybrid tree which is the only one of its kind known; its seeds have been distributed all over the United States. Geneva lies in the heart of Limberlost Swamp country, immortalized by the writings of Gene Stratton Porter; Limberlost State Me-

New friends join to make a castle. (Corps of Engineers photo)

morial contains the cabin in which she lived for 20 years. The campus of Concordia Senior College near Fort Wayne, the only accredited two-year-senior college in the United States, has received international recognition for its design. Peru, one-time wintering spot for many circuses, has a museum reflecting its heritage and an outstanding Circus Festival each July featuring child stars. Noted composer Cole Porter was born here and is buried in Mount Hope Cemetery. Wabash was the first electrically lighted city in the world; at the courthouse you can see one of the first electric lamps, as well as one of the best statues of Abraham Lincoln ever created. The Billman Monument Company at Logansport has some fascinating carving exhibits.

FOR ADDITIONAL INFORMATION:

Public Affairs Officer
U.S. Army Engineer District, Louisville
P.O. Box 59
Louisville, KY 40201

Manager
Salamonie Reservoir
Route 7, Box 99
Huntington, IN 46750

ADDITIONAL CORPS OF ENGINEERS LAKES*

Cagles Mill Lake (from Terre Haute, 30 mi. E on I-70, 5 mi. S on SR 243); Huntington Lake§ (S of Huntington on SR 5); Mississinewa Lake (from Peru, S on SR 21, E on SR 400, N on 550)

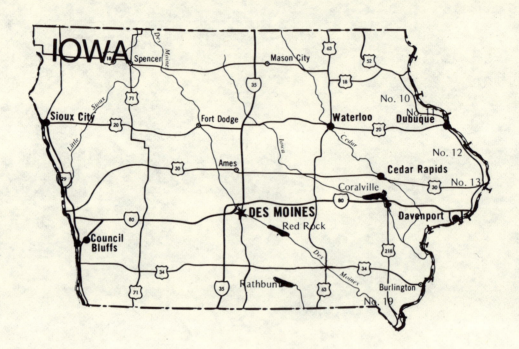

IOWA

CORALVILLE LAKE

Coralville Lake, surrounded by rolling meadows and wooded hills, lies in the east central part of Iowa. Restrained by the narrow valleys of the Iowa River and its tributary streams, the reservoir creates 68 miles of irregular shoreline along its 22-mile path. Lake Macbride, a 935-acre subimpoundment of Coralville Lake, is bordered by a state park. Facilities on the main reservoir are managed primarily by the Corps of Engineers.

HOW TO GET THERE: From Iowa City, go north on I-80 to Dubuque St. interchange, then north again on North Liberty Rd.

FISHING: Crappie is a major fish species here. Northern pike, white bass, and bullheads have also pleased many an angler. Other species available in-

*Unless designated as follows, each project has restrooms, drinking water, and developed campsites. † = no restrooms; ‡ = no drinking water; § = no developed campsites.

clude channel catfish, flatheads, walleye, bass, and bluegill. Inland commercial netting is permitted here to reduce the rough fish population; commercial fishermen run their nets to remove buffalo, carp, carpsuckers, and drum. About a dozen launching ramps; lakeside marina, rental boats, supplies.

HUNTING: The Hawkeye Wildlife Area has been established on over 13,000 acres of reservoir land as a waterfowl sanctuary and hunting area. The ring-necked pheasant is the most highly prized game here. Quail, rabbit, squirrel, and white-tailed deer provide additional action. Though some wild turkey has been stocked here, a turkey season is not yet available. Waterfowl species are snow, blue, and Canada geese and mallard ducks.

CAMPING: Around 350 developed Corps of Engineers campsites; facilities include restrooms, drinking water, showers, picnic tables, grills, dump stations, and launching ramps. Lake Macbride State Park provides 200 additional campsites, both primitive and modern, some with electricity.

OTHER ACTIVITIES: Lake swimming, water skiing, boating of all types; lakeside marina, rental boats, about a dozen launching ramps. Picnic areas and hiking trails around lake. Good spot for rockhounding. The state park offers golf and a softball diamond,

as well as snowmobiling when weather permits. Hawkeye Wildlife Sanctuary, which occupies over 13,000 acres of land and water in the upper reaches of the project, is a haven for nature lovers and bird-watchers.

AREA ATTRACTIONS: West Branch is the birthplace of President Herbert Hoover; visitors may see his home, library, and gravesite at this National Historical Site. The seven Amana Colonies northwest of Iowa City are world-famous for their food, wines, furniture, and woolens; be sure to sample the home-style cooking in their restaurants and tour their industries. At Cedar Rapids are nearly 60 city parks, which offer such attractions as a zoo, golf courses, swimming, and winter sports. Noted artist Grant Wood lived in Cedar Rapids at one time; he designed the large stained-glass Veterans Memorial Window in the Memorial Coliseum. The M. A. Stainbrook State Preserve near North Liberty is rich in fossils, while the Williams Prairie State Preserve near Oxford abounds with rare plants.

FOR ADDITIONAL INFORMATION:
Public Affairs Officer
U.S. Army Engineer District, Rock Island
Clock Tower Building
Rock Island, IL 61201

RATHBUN LAKE

Extending 11 miles up the Chariton River Valley, Rathbun Lake in south central Iowa offers 11,000 sprawling acres to outdoor recreation lovers. An extensive forestation program is underway here to supplement the bottomland hardwoods which now occur along the river and tributary streambanks. This is the Hawkeye State's largest and most popular Corps of Engineers reservoir. Facilities are operated by the Corps and the state of Iowa.

HOW TO GET THERE: From Ottumwa, head west on U.S. 34 for 21 mi. to SR 5; turn south for about 16 mi. to dam.

FISHING: Anglers vie for channel catfish, walleye, striped bass, largemouth bass, white bass, and muskie. Nearly a dozen launching ramps; lakeside marina, rental boats, supplies.

HUNTING: White-tailed deer numbers are good, but hunting pressure is heavy. One of three turkey hunting

zones in the state is in this area. Other game includes quail, pheasant, rabbit, squirrel, and waterfowl. About 20,000 acres of reservoir land are managed by the Corps of Engineers for public hunting.

CAMPING: Over 500 Corps of Engineers developed campsites; facilities include restrooms, showers, drinking water, tables, grills, dump stations, and launching ramps. Honey Creek State Park at the reservoir's edge offers another 200 modern campsites, some with electricity.

OTHER ACTIVITIES: Swimming, water skiing, boating of all types; lakeside marina, rental boats, nearly a dozen launching ramps. Picnic areas around reservoir. Hiking in Honey Creek State Park, as well as some Indian mounds to explore.

AREA ATTRACTIONS: The Wayne County Historical Museum is at Corydon. Sharon Bluffs State Park near Centerville features a scenic overlook atop Chariton River bluffs. The Dutch community of Pella has a Historical Restoration Site, which includes the boyhood home of famous western marshal Wyatt Earp and a craft shop where wooden shoes are made and sold.

FOR ADDITIONAL INFORMATION:
Public Affairs Officer
U.S. Army Engineer District, Kansas City
700 Federal Bldg.
601 E. 12th St.
Kansas City, MO 64106

ADDITIONAL CORPS OF ENGINEERS LAKES

Lake Red Rock* (4 mi. SW of Pella on County Rd. T15)

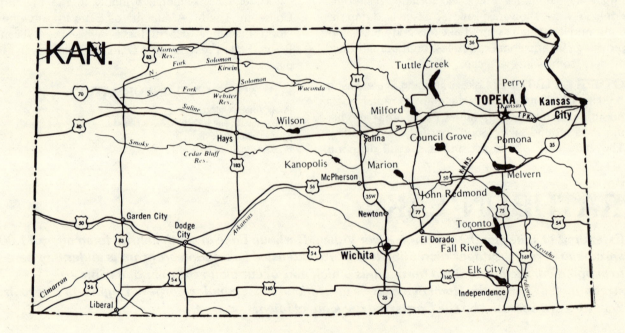

KANSAS

PERRY LAKE

A beautiful, 12,600-acre reservoir located in the rolling hills of northeastern Kansas and fed by the Delaware River, Perry Lake offers a 160-mile shoreline etched with pockets and coves. Daniel

*This project has restrooms, drinking water, and developed campsites.

Boone's third son, Daniel, established the first settlement in the Sunflower State about 5 miles southeast of the town of Perry; Daniel's son, who arrived in 1828, was believed to be the first white child born in what is now Kansas. The reservoir is skirted by an excellent highway system, and both Topeka and Kansas City are within easy driving distance. In addition to Corps of Engineers recreation areas, Perry State Park edges the lake.

HOW TO GET THERE: From Topeka, take U.S. 75 north to U.S. 24. Turn east for about 15 mi. to Perry; turn 3 mi. north to dam.

FISHING: Good fishing for walleye, largemouth bass, crappie, white bass, bullheads, northern pike, and bluegill in the reservoir. The Delaware River has long been noted as one of the top channel cat streams in the state. Nearly a dozen launching ramps; lakeside marinas, rental boats, supplies.

HUNTING: Fox, gray squirrel, raccoon, and coyotes are plentiful; deer, pheasant, and quail populations are fair and on the upswing. Nesting mallards and teal make for excellent duck hunting. Nearly 11,000 acres of game management area.

CAMPING: Nearly 450 developed Corps of Engineers campsites; facilities include restrooms, showers, picnic tables, grills, dump stations, drinking water, and launching ramps. Around 200 modern campsites in Perry State Park.

OTHER ACTIVITIES: Lake swimming, water skiing, boating; lakeside marinas, rental boats, nearly a dozen launching ramps. Sailboating is popular here; also hiking and picnicking. Reservoir lands are located on the old Delaware Indian Reservation, and it's possible some artifacts might be turned up.

AREA ATTRACTIONS: The state capitol building in Topeka is famous for its wall murals. Located on the University of Kansas campus at Lawrence are museums of art, entomology, and natural history. In the same city is the Haskell Indian Institute. The Agricultural Hall of Fame and National Center at Bonner Springs, housed in three buildings, is a monument to our nation's largest industry. The museum at Fort Leavenworth Army Post boasts an outstanding exhibit of nineteenth-century horse-drawn vehicles; also of interest here are the Post Chapel, a National Cemetery, and markers of the Santa Fe and Oregon trails. About 20 miles west of the reservoir is the Potawatomie Indian Reservation.

FOR ADDITIONAL INFORMATION:
Public Affairs Officer
U.S. Army Engineer District, Kansas City
700 Federal Bldg.
601 E. 12th St.
Kansas City, MO 64106
Perry State Park
Meriden, KS 66512

TORONTO RESERVOIR

Blessed with several hundred acres of hardwoods reaching from rolling hills down to the clear water of its 51-mile shoreline, Toronto Reservoir is nestled in the fertile and scenic Verdigris River Valley of southeastern Kansas. The only granite deposit in the Sunflower State was discovered about four miles east of the town of Toronto during the 1880s, and the resulting short-lived boom left the region marked with several deep pits which can still be seen today. This same area also yields the only quartzite in the state. The modern-day rush to this corner of Kansas is for the recreation provided by 2,800-acre Toronto Reservoir, where facilities are managed by the Corps of Engineers and the Kansas State Park Authority.

HOW TO GET THERE: From El Dorado, go 50 mi. east on U.S. 54, then 9 mi. south on SR 105.

FISHING: Kansas' first world record fish was a white bass which came from Toronto Reservoir in 1966, and this same species has received top billing here ever since. Other sport fish found here include black and white crappie, largemouth bass, channel and flathead catfish, bluegill, walleye, and freshwater drum. Nearly half a dozen launching ramps; lakeside marina.

HUNTING: Duck Island near the dam and much of the upper half of the reservoir (both land and water areas) are managed for public hunting. Principal wildlife species available here are bobwhite quail, squirrel, cottontail rabbit, white-tailed deer, mourning dove, ducks, geese, and greater prairie chicken.

CAMPING: About 20 Corps of Engineers campsites near the dam, with over 215 more sites available on Toronto State Park land. Facilities include restrooms, drinking water, showers, tables, grills, dump stations, some electrical hookups (at state park), and launching ramps. Additional public sites available at nearby Wilson and Woodson state parks. Private campgrounds in area.

OTHER ACTIVITIES: Swimming, water skiing, boating; lakeside marina, nearly half a dozen launching ramps. Picnic areas are numerous along the lakeshore. Portions of the Verdigris River good for canoeing.

AREA ATTRACTIONS: A pictorial exhibit of cannibal life, as well as artifacts and photographs from Africa and the South Seas, are housed in the Safari Museum at Chanute; they're representative of the expeditions made by Martin and Osa Johnson, internationally famous adventurers (Osa lived here as a girl). The Flint Hills National Wildlife Refuge is near Burlington; it's located within the flood pool area of another Corps of Engineers' reservoir—John Redmond. Emporia was made famous by the late journalist, William Allen White, who published a newspaper here; his home may be visited, and his papers and writings may be viewed at a library on the Kansas State College campus. Frederick Funston, a great American general of the past who was the hero of the Philippine campaign of 1901, grew up in Iola; his home is now a memorial and museum.

FOR ADDITIONAL INFORMATION:
Public Affairs Officer
U.S. Army Engineer District, Tulsa
P.O. Box 61
Tulsa, OK 74102

Resident Engineer
U.S. Army Engineer Office
P.O. Box 37
Fall River, KS 67047

WILSON LAKE

Nestled in the lush, rolling hills of north central Kansas is one of the Sunflower State's loveliest reservoirs—Wilson Lake. Its 100 miles of boulder-strewn shoreline encircle 9,000 acres of clear blue water. This area of Kansas is noted for its interesting rock formations, created by the weathering of sandstone. The Plains Indians of old exhibited their artistic ability through sandstone "rock carvings" which have survived the ages and may be seen today on reservoir lands. Visitors here are approximately 25 miles from the geodetic center of the North American continent; it's the base for all surveys and mapping in the United States. Bordering the lake is Wilson State Park.

HOW TO GET THERE: From Salina, take I-70 west for 48 mi. to Wilson exit; turn north for 7 mi.

FISHING: Experimentally stocked striped bass have really thrived at Wilson Lake. Well known for its excellent crappie, white bass, and walleye fishing, as well as channel and flathead catfish, largemouth bass, and bluegill. The Cedar Creek area provides good ice fishing during winter months. Half a dozen launching ramps; lakeside marina, rental boats, supplies.

HUNTING: Good populations of quail, pheasant, dove, rabbit, deer, and waterfowl. Present in lesser abundance are squirrel and raccoon. Over 7,000 acres of public hunting land.

CAMPING: Around 150 developed Corps of Engineers campsites; facilities include restrooms, showers, drinking water, tables, grills, dump station, and launching ramps. The Otoe Park Area features a floating dump station for boats. Nearly 200 more campsites at Wilson State Park, some with electric hookups.

OTHER ACTIVITIES: Lake swimming, water skiing, pleasure boating; lakeside marina, rental boats, half a dozen launching ramps. Sailing is becoming increasingly popular here, as it is at all Corps of Engineers reservoirs in Kansas. Lots of shoreline along which to enjoy a picnic lunch or take a hike. Try some rockhounding.

AREA ATTRACTIONS: Custer's famous Seventh Cavalry was once stationed at Fort Hays near the town of Hays; a blockhouse and guardhouse still stand. At the Fort Hays entrance is a large stone carving which memorializes the huge herds of buffalo which once roamed these plains. Russell has a fossil museum. St.

Fidelis Church at Victoria, a remarkable feat of architectural beauty which has stood here since 1911, is known as the "Cathedral of the Plains."

FOR ADDITIONAL INFORMATION:
Public Affairs Officer
U.S. Army Engineer District, Kansas City
700 Federal Bldg.
601 E. 12th St.
Kansas City, MO 64106
Wilson State Park
Lucas, KS 67648

ADDITIONAL CORPS OF ENGINEERS LAKES*

Council Grove Lake (from Topeka, 30 mi. S on I-35, 26 mi. W on U.S. 56); Elk City Lake (from Independence, 7 mi. N on U.S. 75, 4 mi. W and 2 mi. S on county rd.); Fall River Lake (from El Dorado, 10 mi. S on U.S. 77, 46 mi. E on SR 96); John Redmond Reservoir (from Emporia, 27 mi. E on U.S. 50, 25 mi. S on U.S. 75); Kanopolis Lake (from Salina, 26 mi. SE on SR 140, 10 mi. S on SR 141); Marion Lake (from Newton, 24 mi. N on SR 15, 6 mi. E on U.S. 56); Melvern Lake (from Kansas City, 72 mi. SW on I-35); Milford Lake (from Topeka, 63 mi. W on I-70, 5 mi. N on U.S. 77); Pomona Lake (from Topeka, 29 mi. S on U.S. 75, 4 mi. E on SR 268); Tuttle Creek Lake (from Topeka, 45 mi. W on I-70; 15 mi. N on SR 177)

KENTUCKY

DEWEY LAKE

Nestled in the colorful Big Sandy Valley amid the mountains of eastern Kentucky, 1,100-acre Dewey Lake is edged by wooded hills which rise some 700 feet. So magnificent is the scenery that half of the region's 1.5 million annual visitors come primarily to view and explore the natural beauty of this remote area. An impoundment of Johns Creek, Dewey Lake offers all the solitude any visitor desires, as well as an array of outdoor sports and recreation. Much of the attraction of this area lies in the fact that lakeside development has thus far been kept to a minimum. A limited number of Corps of Engineers recreation areas are complemented by the facilities at Jenny Wiley State Resort Park, one of Kentucky's most popular; among them are a lodge, dining room, and cottages.

HOW TO GET THERE: From Prestonsburg, KY, head 3 mi. south on U.S. 23, then 1 mi. east on SR 304.

FISHING: Crappie and white bass lure anglers during spring and summer months. Also good for largemouth bass, bluegill, and catfish. Some muskie have been

---

*These projects all have restrooms, drinking water, and developed campsites.

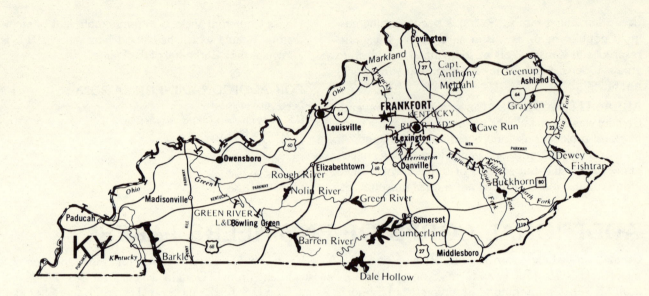

stocked here. Nearly half a dozen launching ramps; lakeside marina, rental boats, supplies.

HUNTING: The extremely steep and rugged, completely forested terrain offers excellent deer hunting. Also grouse, squirrel, fox, quail, rabbit, and raccoon.

CAMPING: Nearly 160 Corps of Engineers campsites, both primitive and developed. Facilities include restrooms, drinking water, tables, grills, and launching ramps. Two Corps recreation areas offer over 125 boat-in campsites only. Additional campsites with full hookups at Jenny Wiley State Park. The city of Prestonsburg operates a modern campground at Archer Park one mile west of town.

OTHER ACTIVITIES: Swimming at white sand beach and in state park pool; water skiing, boating; nearly half a dozen launching ramps, rental boats, lakeside marina. Good canoeing streams in area. Excellent hiking trails. Over 30 shoreline picnic areas. Mountain climbing, horseback riding, nine-hole golf at state park. SR 201 north of the reservoir is a particularly scenic drive. A skylift transports visitors to the top of a mountain for a spectacular view. Outdoor dramas are presented during the summer in the state park's amphitheater.

AREA ATTRACTIONS: Dewey Lake State Forest adjoins the reservoir. Garfield Place in Prestonsburg was the Civil War headquarters of General James A. Garfield, later president of the United States. The grave of Jenny Wiley, for whom the state park was named, is near Paintsville; she was a pioneer woman whose bravery is legendary in these parts. A little over an hour's drive to the west is Daniel Boone National Forest, home of famous Red River Gorge. The unique Breaks Interstate Park, cosponsored by Kentucky and Virginia, lies an hour's drive to the southeast; it's perched atop the rim of the largest river canyon east of the Mississippi River.

FOR ADDITIONAL INFORMATION:

Public Affairs Officer
U.S. Army Engineer District, Huntington
P.O. Box 2127
Huntington, WV 25721

Jenny Wiley State Resort Park
Prestonsburg, KY 41653

LAKE BARKLEY

Twisting like a giant snake within the confines of the narrow Cumberland River plain, Lake Barkley extends from Barkley Dam near Grand Rivers, KY, deep into Tennessee, a distance of nearly 120 miles. This reservoir, Kentucky Lake eight miles to the west, and the unique Land

The beautiful redwood and glass lodge at Barkley State Park.

Between the Lakes (LBL) which separates them form the largest inland water recreation center in the nation—and one of the most popular. Close to 5 million people annually visit this western Kentucky reservoir to enjoy the unpolluted waters, unspoiled woodlands, and vast panorama of outdoor recreational facilities. Along the eastern shoreline of Lake Barkley are over 25 Corps of Engineers recreation areas, as well as the finest resorts, motels, and restaurants. Land Between the Lakes, on the opposite shore, has been developed by the Tennessee Valley Authority (TVA) as a national demonstration in outdoor recreation and environmental education. Perhaps the most notable lodge on the lake is the huge redwood and glass complex at Lake Barkley State Resort Park; there also are housekeeping cottages here.

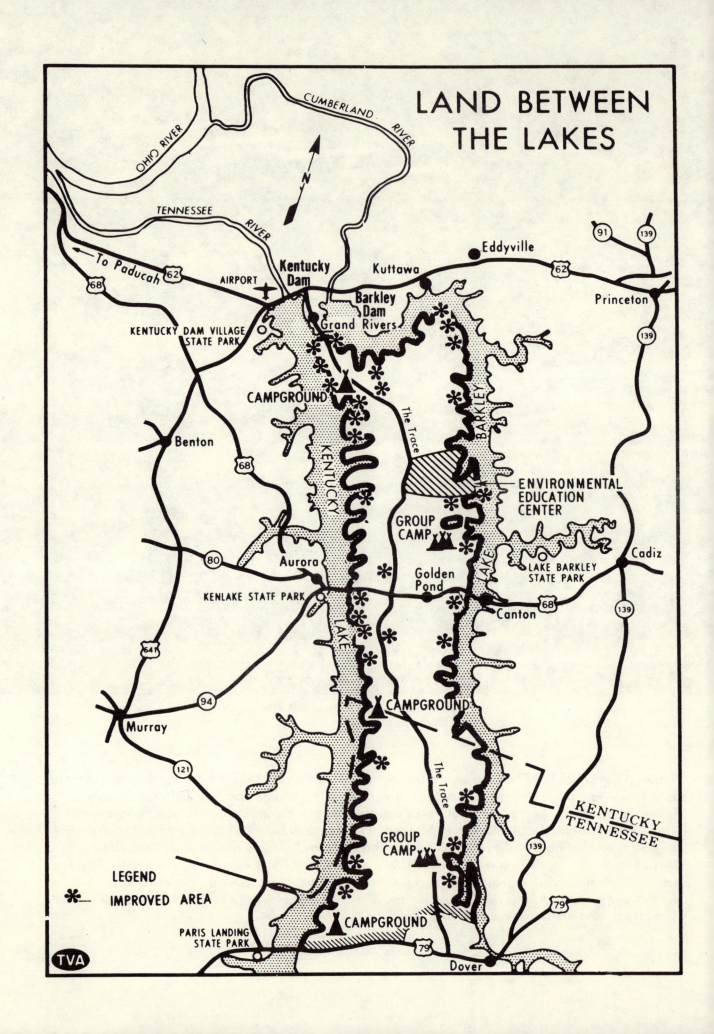

HOW TO GET THERE: From Paducah, KY, head east for approximately 25 mi. on U.S. 62 to Barkley Dam.

FISHING: Many anglers consider Lake Barkley the ideal bass lake, with extensive stump rows and inundated brush piles providing ideal habitat. The spring crappie run has attracted national attention. Also found here are bluegill, stripes, and sauger. Catfish and rockfish inhabit the tailwaters below the dam. Night fishing with lights is very productive during summer months. A short, free-flowing, deep-water canal which connects Lake Barkley and Kentucky Lake near the two dams at the northern ends of the lakes has become a good fishing hole in its own right. With 58,000 acres of water and 1,000-plus miles of shoreline, Lake Barkley offers plenty of hideaway spots. Nearly 50 Corps of Engineers launching ramps; lakeside marinas all along the reservoir's eastern shores (there are no commercial facilities at LBL), rental boats, guides, and supplies.

HUNTING: Primarily on the islands of Lake Barkley (one is maintained as a waterfowl refuge area) and in LBL. Duck and geese hunting is excellent. The turkey at LBL are reputed to be the only remaining flock of true wild turkeys in the United States; very limited season at present. Other game includes deer, squirrel, dove, quail, rabbit, raccoon, and opossum. For specific regulations for LBL, contact TVA; remainder of hunting lands are governed by state laws.

CAMPING: Over 200 Corps of Engineers campsites, all developed; facilities include restrooms, showers, drinking water, tables, grills, playgrounds, dump stations, and launching ramps. LBL campgrounds range from primitive to all-electric travel trailer campgrounds. There are over 800 well-equipped sites in three big LBL family campgrounds and many more at specialty campgrounds and lake access areas all around the shoreline. Some sites are free, and there are special facilities for campers on horseback and the handicapped. Lake Barkley State Resort Park offers 75 modern campsites with full hookups, and numerous private campgrounds are located along the eastern shoreline.

OTHER ACTIVITIES: Swimming beaches at over half a dozen Corps of Engineers recreation areas and at the state park, water skiing, boating; nearly 50 Corps launching ramps, lakeside marinas, and rental boats of all types, including sailboats, houseboats, and canoes. Picnic areas in scenic settings with lake view. Numerous hiking trails on both shores, including one in LBL which accommodates wheelchairs; horseback riding and bicycling. Lake Barkley State Resort Park offers a trapshooting range; basketball, tennis, and shuffleboard courts, and an 18-hole golf course. Many opportunities for wildlife observation and bird-watching, particularly in LBL, where 85 percent of the peninsula's 170,000 acres are heavily forested; bald and golden eagles winter in the area, and a herd of buffalo roams a 200-acre pasture. Some special features of LBL are an archery range, an Off Road Vehicle Area, and many visitor-participation programs.

AREA ATTRACTIONS: Visitors to the two dams will have the opportunity to observe huge barges and pleasure boats as they pass through the massive locks. LBL's Environmental Education Center has an interpretive center, an educational farm, and an overlook atop a silo. Fort Donelson near Dover, TN, is a national military park. There are several large tobacco markets in the area; winter visitors may want to attend the auctions.

FOR ADDITIONAL INFORMATION:

Public Affairs Officer
U.S. Army Engineer District, Nashville
P.O. Box 1070
Nashville, TN 37202

Resource Manager
Barkley Dam
P.O. Box 218
Grand Rivers, KY 42045

Land Between the Lakes
Tennessee Valley Authority
Golden Pond, KY 42231

Lake Barkley State Resort Park
Cadiz, KY 42211

LAKE CUMBERLAND

In the winding, twisting valley of the upper Cumberland River, stretching over 100 miles from Wolf Creek Dam into Daniel Boone National Forest, lies lovely Lake Cumberland. The 50,000-acre reservoir, dotted with islands and surrounded by an irregular shoreline of cliffs and the tree-

covered foothills of the Cumberland Mountains, is a seeming paradox. Some of the nation's finest and most modern vacation facilities are along its shores (including two lodges and numerous cottages at Lake Cumberland State Park); yet much of the land surrounding these clear, deep waters seems as rugged and pristine as it might have been when Daniel Boone first passed this way. Located in sparsely populated southeastern Kentucky, Lake Cumberland is one of the most popular Corps of Engineers impoundments in the country. In addition to several Corps recreation areas, two state parks, a county park, and some private enterprises provide services and facilities for nearly 5 million annual visitors.

HOW TO GET THERE: From Russell Springs, take U.S. 127 south for about 20 mi. to the dam.

FISHING: Reputed to be one of the best fishing areas in Kentucky, Lake Cumberland has produced several state records. Anglers will find largemouth, small-mouth, Kentucky, and white bass, crappie, bluegill, rockfish, and walleye. A record rainbow trout was taken from the tailwaters below the dam. The reservoir is fished year around, but is nationally famous for its spring run of white bass. Best black bass fishing is in the fall. Deep, exceptionally clear waters encourage spearfishing. Lots of secluded fishing holes. Nearly a dozen Corps of Engineers launching ramps; lakeside marinas, rental boats, supplies.

HUNTING: Wide areas of virtually untouched wilderness offer deer, squirrel, rabbit, fox, raccoon, and quail. Wild turkey is making a comeback in Daniel Boone National Forest.

CAMPING: About 150 developed campsites in Corps of Engineers recreation areas around the lake. Facilities include restrooms, drinking water, picnic tables and grills, showers, and boat launching ramps. Camping also in Lake Cumberland State Resort Park and General Burnside Island State Park, as well as Pulaski County Park and numerous private campgrounds; wide range of facilities. Boat camping along reservoir shores and on upper Lake Cumberland in the Daniel Boone National Forest.

OTHER ACTIVITIES: Lake swimming at several beaches or in pool at Lake Cumberland State Park; clear, deep water lures skin and scuba divers. Water skiing and boating of all types extremely popular; nearly a dozen launching ramps, lakeside marinas, rental boats (including houseboats). Also picnicking, hiking, horseback riding. Burnside Island State Park

has an 18-hole golf course; Lake Cumberland State Resort Park has a 9-hole golf course, as well as tennis and shuffleboard courts. Excellent canoeing, boating, and backpacking opportunities in Daniel Boone National Forest.

AREA ATTRACTIONS: Many scenic waterfalls in area; notable ones are Seventy-six Falls on the reservoir's Indian Creek and Cumberland Falls in the nearby state park of the same name. The latter is one of two falls in the world which reflect a moonbow (night rainbow) on nights of the full moon. Numerous scenic drives; one of the loveliest is Red River Gorge Loop Drive in Daniel Boone National Forest. During summers, boaters may attend church without leaving the water; the minister remains on dry land near Monticello, leading the service over a loudspeaker. At Mill Springs is the largest working water wheel in the world, and you can buy water-ground cornmeal there. Just north of Burkesville is the site of what was reputed to be the first American oil well. At Renfro Valley, Saturday night barn dances and Sunday morning hymn sings have been a tradition for years.

FOR ADDITIONAL INFORMATION:
Public Affairs Officer
U.S. Army Engineer District, Nashville
P.O. Box 1070
Nashville, TN 37202
Reservoir Manager
Lake Cumberland
Jamestown, KY 42629
Forest Supervisor
Daniel Boone National Forest
100 Vaught Rd.
Winchester, KY 40391
Lake Cumberland State Resort Park
Jamestown, KY 42629
General Burnside Island State Park
Burnside, KY 42519

ROUGH RIVER LAKE

In or out of the water, visitors will find plenty to enjoy at 3,000-acre Rough River Lake, located among the gentle hills of west central Kentucky. The rough terrain and high banks along much of the shoreline are lushly wooded, a sanctuary for nature lovers, while the areas surrounding the public lands are heavily developed and offer a vast variety of recreational opportunities. This combination attracts over 1.25 million visitors yearly. A state resort park near the dam features a rustic lodge, housekeeping cottages, and its own airstrip.

HOW TO GET THERE: From Louisville, take U.S. 60 west for about 65 mi. to Harned; turn south on SR 79 for 10 mi.

FISHING: One of the largest populations of walleye in the state, ranging up to 26 inches in size, is found here. Also a good lake for black and white bass, crappie, and bluegill, especially in the spring. A rainbow trout hatchery has been established below the dam, and the trout supply is replenished monthly spring through fall. Over half a dozen launching ramps; marina, rental boats, and supplies at lakeside.

HUNTING: Deer, squirrel, rabbit, and quail inhabit the woods around Rough River Lake. Hunters are urged to use care because areas adjoining the public lands are so heavily developed.

CAMPING: Nearly 500 Corps of Engineers campsites, both primitive and developed. Facilities include picnic tables and grills, restrooms, drinking water, showers, dump stations, and launching ramps. Additional campsites at Rough River Dam State Resort Park.

OTHER ACTIVITIES: Beach swimming, water skiing, and boating; lakeside marina, rental boats, over half a dozen launching ramps. Along the shoreline are several picnic grounds. Also horseback riding, golf on a nine-hole course, tennis, hiking trails, children's playground, and a summer theater, all at the state park.

AREA ATTRACTIONS: For sightseers and history buffs, the Abraham Lincoln Trail passes through Madrid just north of the reservoir; it's ideal for bicycle touring as well as motorists. The trail begins at the Lincoln Birthplace National Historic Site near Hodgenville, approximately an hour's drive to the east. Elizabethtown offers a tour of 14 historical homes; among them is the former residence of Lt. Col. George Custer. Fort Knox Military Reservation is just north of Elizabethtown on U.S. 31W; the Patton Museum is open to the public. About 45 miles southeast is Mammoth Cave National Park. Owensboro on the Ohio River to the northwest has many attractions, including a tour of Glenmore Distillery and Museum. Another tour, as well as a sampling room, is offered at Albany Cheese, Inc., near Leitchfield.

FOR ADDITIONAL INFORMATION:

Public Affairs Officer
U.S. Army Engineer District, Louisville
P.O. Box 59
Louisville, KY 40201

Reservoir Manager
Rough River Lake
Falls of Rough, KY 40119

Rough River Dam State Resort Park
Falls of Rough, KY 40119

ADDITIONAL CORPS OF ENGINEERS LAKES*

Barren River Lake (from Louisville, 95 mi. S on I-65 to Cave City, 10 mi. S on SR 90 to Glasgow, 5 mi. S on U.S. 31 to SR 252, 9 mi. S to dam); Buckhorn Lake (from Hazard, 9 mi. N on SR 15, 20 mi. E on SR 28); Fishtrap Lake (from Pikeville, 12 mi. E on U.S. 460, 2 mi. E on SR 1789); Grayson Lake (7 mi. S of Grayson on SR 7); Green River Lake (90 mi. SE of Louisville, via I-65 and SR 61, 210, and 55); Markland Locks and Dam§ (from Louisville, 50 mi. E on I-71, 5 mi. N on SR 227, 14 mi. E on SR 42 to Markland); Nolin River Lake (from Brownsville, N 5 mi. on SR 259, right on SR 728, follow signs)

---

*Unless designated as follows, each project has restrooms, drinking water, and developed campsites. † = no restrooms; ‡ = no drinking water; § = no developed campsites.

Proud fishermen display a nice catch from Lake Barkley.

LOUISIANA

BAYOU BODCAU

Bayou Bodcau Dam, named for the stream it impounds, is located about 23 miles northeast of Shreveport, LA. The project was originally designed for flood control; and, although the lake will cover 21,000 acres and extend 30 miles upstream into Arkansas during high-water periods, there is no permanent pool here. Therefore, recreational facilities on the reservoir are limited. Good hunting and fishing, however, have contributed to a rise in visitation from an estimated 8,000 people in 1955 to nearly 332,000 in 1974. In addition to Bodcau Lake, over half a dozen cypress-studded lakes of haunting beauty attract sightseers and sportsmen to this northeastern corner of the Pelican State. Just east of Bayou Bodcau lies a division of Kisatchie National Forest, and here the terrain is dominated by pine forests. Lake Ivan is permanently maintained by Bossier Parish (County) on 520 acres of reservoir lands and provides additional recreational opportunities. Nearby Shreveport offers the finest of modern accommodations, and there are numerous private and public campgrounds in the area.

HOW TO GET THERE: From Shreveport, head east on I-20 for 20 miles; then turn north on SR 157 for eight miles to Bellevue and follow marked county road another two miles north.

FISHING: Bass is the big drawing card here, primarily largemouth and smallmouth. Other popular species are crappie, bream, and catfish. Fishing is best throughout the summer months. About five launching ramps.

HUNTING: The Bodcau Wildlife Management Area represents nearly 33,000 acres of rolling pine and hardwood lands stretching southward from Springhill for some 25 miles to near Bellevue. The land is owned by the Corps of Engineers. Game species hunted, in order of importance, are ducks, deer, squirrel, woodcock, quail, and dove. Wild turkeys have been released on the area and increase in population each year. Additional hunting opportunities in Kisatchie National Forest.

CAMPING: The Corps of Engineers offers a limited number of primitive campsites (around 10) near the dam; facilities include restrooms, drinking water, and tables. There are plenty of public and private campsites within short driving distance, however. Caney Lakes Recreation Area, northwest of Minden in Kisatchie National Forest, offers modern restrooms and showers as well as a lakeside setting. Two state parks with campsites and water-based recreation, Lake Claiborne State Park and Lake Bisteneau State Park, are both within an hour's drive of Bodcau Lake. And the Shreveport-Bossier KOA is a private campground with full hookups, swimming pool, and playgrounds.

OTHER ACTIVITIES: Lake swimming and boating. Water skiing on Lake Ivan and Caney Lakes. About five launching ramps. Canoeing on the area's many bayous, including Bodcau. Picnic areas near Bayou Bodcau Dam. Hiking and bird-watching (more than 250 species) in Kisatchie National Forest.

AREA ATTRACTIONS: The Louisiana State Exhibit Museum in Shreveport has achieved international recognition for dioramas, rotating art shows, and special exhibits. That same city's unique redevelopment area, Shreve Square, has been described as a mini-New Orleans and underground Atlanta combined. Barksdale Air Force Base in Bossier City is one of the largest air bases in the world. If you're in the area in the early spring, be sure to take the Dogwood Trail Drive north from the town of Plain Dealing along SR 157; it winds through the highest hills in the state and is enhanced by dogwood, redbud, and wild flowers in bloom.

FOR ADDITIONAL INFORMATION:

Public Affairs Officer
U.S. Army Engineer District, New Orleans
P.O. Box 60267
New Orleans, LA 70160

Forest Supervisor
Kisatchie National Forest
2500 Shreveport Hwy.
Pineville, LA 71360

Reservoir Superintendent
Bayou Bocau Reservoir
P.O. Box 1817
Texarkana, TX 75501

ADDITIONAL CORPS OF ENGINEERS LAKES

Pearl River Lock No. 1* (from Hattiesburg, MS, 76 mi. S on I-59, 12 mi. N on SR 41 to access road)

MASSACHUSETTS

BIRCH HILL RESERVOIR

In the north central section of Massachusetts is one of New England's most popular lakeside recreation areas—Birch Hill Reservoir. The core of recreational activities is an 82-acre natural

*This project has no restrooms, drinking water, or developed campsites.

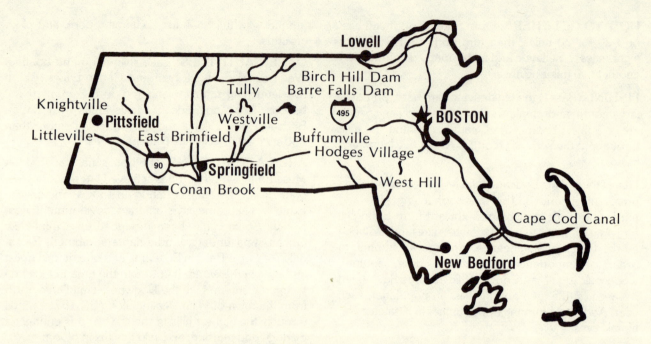

lake, Lake Denison, surrounded by wild upland forest on a gently rolling terrain. Most facilities are managed by the Massachusetts Department of Natural Resources. Otter River State Forest and Lake Denison State Park are on reservoir lands.

HOW TO GET THERE: From Fitchburg, take SR 2A south to SR 2. Travel west for about 10 mi. to SR 68; head northwest for about 8 mi. to Baldwinsville and pick up project road.

FISHING: A variety of fish in Lake Denison and the adjacent Millers and Otter rivers (which meet nearby) includes trout, pickerel, brown bullhead, largemouth bass, bluegill, yellow perch, black crappie, and pumpkinseed (common sunfish). Two Corps of Engineers launching ramps; an additional half-dozen maintained by the state. Ice fishing.

HUNTING: Sportsmen will find white-tailed deer, gray squirrel, snowshoe hare, cottontail rabbit, quail, ruffed grouse, and pheasant (cocks only may be hunted). Birch Hill State Wildlife Management Area adjoins the lake.

CAMPING: Around 250 developed and primitive campsites; some in Lake Denison State Park, some in Otter River State Forest. Facilities include showers, restrooms, drinking water, tables, grills, and launching ramps.

OTHER ACTIVITIES: Swimming, pleasure boating; about eight launching ramps. A network of scenic trails skirts the reservoir; and there are tables and fireplaces for picnickers. Many scenic drives nearby. Horseback riding is popular. Winter brings snowmobiling, ski touring, and ice skating.

AREA ATTRACTIONS: Leominster is the birthplace of John Chapman, better known as Johnny Appleseed; a plaque along the riverbank marks the area. The Wachusett Meadows Bird Sanctuary is southwest of Fitchburg near Wachusett Mountain. Worcester has an outstanding art museum and the Higgins Armory Museum, which traces the history of steel and exhibits the country's best collection of medieval armory. The Thayer Bird Museum at Lancaster houses a rare ornithological collection. Cook's Canyon is a beautiful scenic spot near Barre. Near Harvard is Fruitlands Museum, comprised of four buildings; one contains memorabilia of Louisa May Alcott, who was born in this city.

FOR ADDITIONAL INFORMATION:

Public Affairs Officer
U.S. Army Engineer Division, New England
424 Trapelo Rd.
Waltham, MA 02154

Manager
Otter River State Forest
Baldwinville, MA 01436

BUFFUMVILLE RESERVOIR

Buffumville Reservoir, lying across the Little River in south central Massachusetts, is one of the Bay State's most popular recreation spots. The low, wooded hills surrounding the lake embrace six miles of scenic shoreline. This is rich historical country, and New Englanders proudly display their heritage in museums and restorations. History buffs will enjoy the old tombstones and the stories they tell in the area's centuries-old graveyards. A 50-acre public park has been developed at Buffumville Reservoir by the Corps of Engineers, with recreation facilities operated by the state's Department of Natural Resources.

HOW TO GET THERE: Take SR 12 south from Worcester for 6 mi.; at Oxford, turn right on Charlton St. for 2 mi.

FISHING: Bass and pickerel fishing is good here, as well as at Hodges Village Dam, another Corps of Engineers reservoir about two miles to the east on the French River. Other species found at Buffumville are yellow perch, bluegill, pumpkinseeds, brown bullhead, black crappie, white sucker, and golden shiners. Ice fishing in winter. Three launching ramps.

HUNTING: Waterfowl and rabbits are the chief quarries for hunters on designated reservoir land.

CAMPING: Although no camping is permitted at either Buffumville or Hodges Village reservoirs, campers may stay at Wells State Park about 25 miles to the west and at numerous private campgrounds within a half-hour's drive.

OTHER ACTIVITIES: Swimming, water skiing, boating; three launching ramps. Several picnic areas. A network of scenic foot trails skirts the reservoir, and snowmobile trails are nearby.

AREA ATTRACTIONS: Visitors may travel to another century by entering Old Sturbridge Village near present-day Sturbridge; one of the finest historical restorations in the country, the village is a living demonstration of New England life as it was in the early 1800s. The art museum in Worcester is one of the finest in the United States. Higgins Armory Museum in the same city has the best collection of medieval armory in the country, as well as exhibits tracing the past and future of the steel industry. The birthplace of Clara Barton, founder of the American Red Cross, may be visited in North Oxford. Just to say they've been there, many people visit a lake southeast of Webster which the Indians gave one of the longest names in the world—Lake Chargoggagogmanchaugagogchaubunagungamaug. Translated it means: "You fish on your side; I fish on mine; nobody fish in the middle." For obvious reasons, it is more often called Lake Webster.

FOR ADDITIONAL INFORMATION:
Public Affairs Officer
U.S. Army Engineer Division, New England
424 Trapelo Rd.
Waltham, MA 02154
Central Massachusetts Tourist Council
90 Madison St.
Worcester, MA 01608

KNIGHTVILLE DAM

Deep in the forested hills of southwestern Massachusetts, the Corps of Engineers has built Knightville Dam across the turbulent Westfield River. Though the reservoir is normally kept empty except during time of flooding, the Westfield River and the dense woods which border it have much to offer in the way of recreation. Sportsmen in particular are attracted to the area. For their use and for the use of the general public, the Corps maintains some recreation facilities here. However, because of much water flow variation, virtually no boating is done here.

HOW TO GET THERE: From Northampton, take SR 66 west for about 15 mi. and turn north on SR 112 to dam road.

FISHING: The river and tributary streams in the reservoir area provide excellent habitat for brown, rainbow, and brook trout. Other species include white

White water canoeing at Knightville Dam. (Corps of Engineers photo)

sucker, brown bullhead, yellow perch, golden and common shiner, creek chub, fallfish, bonded killifish, and pumpkinseed (common sunfish).

HUNTING: The area around Knightville Dam is recognized as the best upland game habitat in western Massachusetts. Hunters will find white-tailed deer, snowshoe hare, cottontail rabbit, squirrel, grouse, quail, and pheasant.

CAMPING: Around 50 Corps of Engineers campsites,

both primitive and developed; at present, facilities include restrooms, tables, and grills. Numerous private campgrounds in area.

OTHER ACTIVITIES: River swimming; hiking trails through forested hills. Many scenic drives in area. The Westfield River is probably the best canoe stream in Massachusetts. With adequate flow, the river from Cummington south may be run by experts. Horseback riding at nearby Gardner State Park, a day-use facility. Ski touring, downhill skiing, snowmobiling, and ice skating are favorite regional winter sports. There's some excellent rockhounding here, particularly along the Connecticut River banks between Northampton and Holyoke.

AREA ATTRACTIONS: Hancock Shaker Village, a restoration of a 1790 community, is near Pittsfield. The evolution of American papermaking may be viewed at Crane Museum near Dalton. Look Memorial Park at Northampton contains many rare shrub plantings; recreational facilities include a swimming pool and tennis courts. American poets Emily Dickinson, Robert Frost, and Eugene Field lived in the college town of Amherst, and special collections of their work may be seen at Jones Library. The Joseph Allen Skinner State Park near South Hadley features panoramic views and interesting rock formations. About five miles north of Holyoke, on the bank of the Connecticut River, is a ledge containing authentic dinosaur footprints; look for a sign marking the site. In Springfield, visitors may see the Armory Museum; featuring weapon displays, and the National Basketball Hall of Fame. The floral gardens and carillon concerts at Westfield's Stanley Park are of interest.

FOR ADDITIONAL INFORMATION:
Public Affairs Officer
U.S. Army Engineer Division, New England
424 Trapelo Rd.
Waltham, MA 02154

ADDITIONAL CORPS OF ENGINEERS LAKES*

Barre Falls Dam§ (from Gardner, S on SR 68 to SR 62, 2 mi. W to dam); Cape Cod Canal (from Plymouth, take SR 3 S to eastern end of canal); Conant Brook Dam†, ‡, § (from Springfield, E on SR 20, S on SR 32); East Brimfield Lake§ (from Worcester, I-290, I-90, and U.S. 20 W); Hodges Village Dam§ (E of Worcester on SR 12 or 20); Littleville Lake§ (from Westville, W on SR 20 and 112); Tully Lake ‡, § (from Boston, SR 2 W to Athol); West Hill Dam§ (from Worcester, 15 mi. S on SR 146 to SR 16, 3 mi. E to signs, N to project); Westville Lake§ (from Southbridge, take Main St. or Rte. 131 to South St.)

MINNESOTA

CROSS LAKE

Encircled by a pine-covered shoreline, Cross Lake in north central Minnesota has one of the finest and most elaborately equipped Corps of Engineers recreation areas in its district. The lake itself is one of the Whitefish chain of lakes, named for Lower Whitefish and Upper Whitefish lakes nearby. From Cross Lake boaters may explore thousands of acres of recreational waters in a dozen or so lakes, all interconnected by rivers and channels. The reservoir lies in the heart of Crow Wing State Forest, and the surrounding countryside is one of the Gopher State's leading resort areas.

HOW TO GET THERE: Take U.S. 210 east from Brainerd. Just outside of town, pick up SR 25 north for 8 mi., turn right on County Rd. 3, and continue north for 18 mi. to the lake.

FISHING: Walleye is the number one choice of most anglers here. Muskie are also popular, but not as abundant. Other fish include northern pike, bass, and panfish. Three launching ramps.

HUNTING: No hunting permitted on reservoir lands, but the vast Chippewa National Forest is not far to the

*Unless designated as follows, each project has restrooms, drinking water, and developed campsites. † = no restrooms; ‡ = no drinking water; § = no developed campsites.

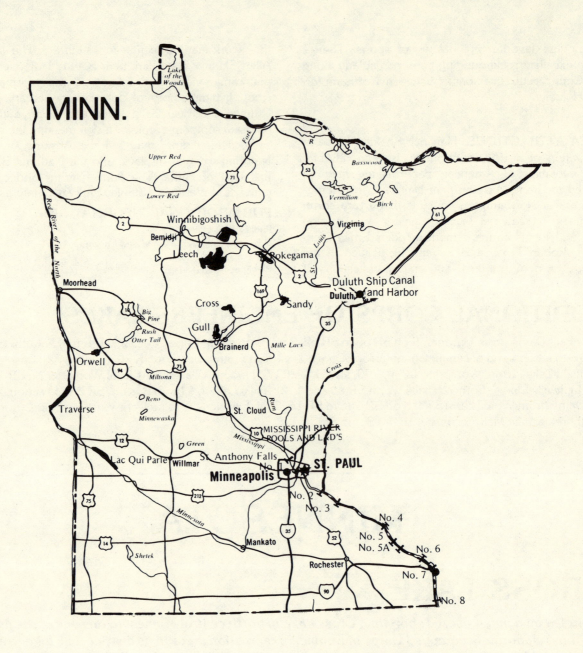

north. Deer and moose are popular big game, while the ruffed grouse is the favorite game bird. Also rabbit, quail, and waterfowl.

CAMPING: Nearly 100 developed Corps of Engineers campsites, with more planned. Facilities include restrooms, drinking water with taps throughout the campsites, showers, tables, grills, firewood, fish cleaning tables, laundry room with tubs, dump stations, and launching ramps.

OTHER ACTIVITIES: Swimming, water skiing, and boating on 14,000-acre Cross Lake, as well as other reservoirs in the interconnected chain of lakes; three launching ramps. Picnic areas around reservoir. Excellent hiking trails in nearby Chippewa National Forest. The Pine River below the dam is a scenic mix of woodland, marsh, and farmland which offers excellent canoeing for those with average skills. There's a 26-mile scenic drive at Camp Ripley Military Reservation southwest of Brainerd. Winter sports are very popular; they include ski touring, snowmobiling, tobogganing, snowshoeing, and ice fishing, all nearby.

AREA ATTRACTIONS: Two of the area's top attractions are near Brainerd; Lumbertown USA is a reconstruction of an 1870 Minnesota village, while the Bunyan Center features a 27-foot, animated likeness of Paul Bunyan which tells tall tales about his legendary exploits. At Rice Lake National Wildlife Refuge near McGregor, visitors may view a great blue heron colony, wild rice beds, and Indian burial mounds. Ak-Sar-Ben near Aitkin features formal gardens and fish which may be fed by hand. Deer may be hand fed at the Deer Forest and Storybook Land near Nisswa, while the Deer Forest Express offers scenic mile-long train rides.

FOR ADDITIONAL INFORMATION:
Public Affairs Officer
U.S. Army Engineer District, St. Paul
1135 U.S. Post Office and Custom House
St. Paul, MN 55101

LEECH LAKE

More than 600 miles of shoreline encompass Leech Lake, the largest resort lake in northern Minnesota. Along the way its waters border Corps of Engineers land, a Chippewa Indian reservation, a national forest, two state forests, and several hundred privately owned resorts. This is the land which spawned the legend of Paul Bunyan, woodsman and lumberjack extraordinary. Chippewa National Forest is one of our country's most important breeding areas for the bald eagle, and one of the areas in which visitors are most likely to see the bird is around the Corps of Engineers dam which impounds Leech Lake. Most recreation facilities here are privately owned, but some facilities are operated by both the Corps and the U.S. Forest Service.

HOW TO GET THERE: From Grand Rapids, head west on U.S. 2, then south on Secondary Rd. 8 to dam, a total of 43 mi.

FISHING: Some muskie history has been made at Leech Lake. A 51-pound specimen taken here in 1973 is on display at Walker's Chamber of Commerce. The number 51 was also significant in setting a different type of record here; that was the total of muskie taken from about one square mile of water during one July weekend back in the 1950s. Walleyes, northern pike, bass, crappie, and bluegill also call these waters home. Kebekona Creek, which empties into Leech Lake from the west, is considered one of the state's finest trout streams. Lakeside marinas, rental boats, supplies; launching ramps.

HUNTING: Deer and waterfowl provide the most outstanding hunting. Also moose, grouse, and snowshoe hare.

CAMPING: Nearly 60 developed Corps of Engineers campsites; facilities include drinking water, tables, grills, restrooms, dump station, and launching ramps. One national forest campground on the lake, and others within the forest nearby. Numerous private campgrounds around the lakeshore and in the vicinity. Boaters who camp can travel around the reservoir from one campground to another; nearly all are accessible by water.

OTHER ACTIVITIES: Swimming, water skiing, skin and scuba diving, pleasure boating of all types, including houseboating and sailing; lakeside marinas, rental boats, launching ramp. The state's largest sailboat regatta is held here over the Fourth of July weekend. There are some prime canoe trails all around. Hikers and picnickers will have no trouble finding a private spot. The national forest offers ample opportunity for nature study; some of the animals who live here are bear, deer, timber wolf, moose, muskrat, and otter. Favorite winter sports are snowmobiling, ski touring, and ice fishing. The Tianna Golf Course at Walker offers a challenging 18 holes.

AREA ATTRACTIONS: A Chippewa Indian burial

An air of helpful interest is shown by a Corps of Engineers Park Ranger. (Corps of Engineers photo)

ground is near the Forest Service's Stony Point campground. Climb a forest lookout tower for a panoramic view. The Museum of Natural History and Indian Arts and Crafts is housed in the Walker Chamber of Commerce Building. There's a monument at Sugar Point on Leech Lake to mark the site of the last Indian battle in the United States for which federal troops were called. The national forest is the home and present headquarters of the Chippewas. Children in particular will enjoy the Deer Valley Deer Farm near Walker, where they'll see tame deer, performing goats, buffalo, and bear. There's a state fish hatchery at Deer River; this town is also the center of a wild rice industry, and the beds are still harvested by primitive methods. Bemidji is the home of the Paul Bunyan Summer Water Carnival and the Paul Bunyan Summer Playhouse. The mighty Mississippi River is born within nearby Itasca State Park; from Lake Itasca it begins its long journey as a shallow, narrow ripple even a child can wade across. Other features of this park are a buffalo pen, daily launch excursions on Lake Itasca, and nearly every kind of wild animal, tree, and plant native to the Gopher State.

FOR ADDITIONAL INFORMATION:
Public Affairs Officer
U.S. Army Engineer District, St. Paul
1135 U.S. Post Office and Custom House
St. Paul, MN 55101
Forest Supervisor
Chippewa National Forest
Cass Lake, MN 56633
Leech Lake Area Chamber of Commerce
Walker, MN 56484

ADDITIONAL CORPS OF ENGINEERS LAKES*

Gull Lake (from Brainerd, 7 mi. N on U.S. 371, 2 mi. W on County 125, 1.5 mi. W on County 105); Lac Qui Parle§ (from Montevideo, 9 mi. NW on SR 7 and 59); Lake Traverse‡, § (from Wheaton, 4 mi. N on U.S. 75, 4 mi. W on SR 236); Lock and Dam No. 3, Mississippi River§ (near Welch, 10 mi. E of U.S. 61); Orwell Lake§ (from Fergus Falls, 10 mi. SW via County Rte. 15); Pokegama Lake (from Grand Rapids, 2 mi. W on U.S. 2); Sandy Lake (from McGregor, 13 mi. N on SR 65); Winnibigoshish Lake (from Deer River, 1 mi. W on U.S. 2, 12 mi. NW on U.S. 46, 2 mi. E on County Hwy 9)

MISSISSIPPI

GRENADA LAKE

This lovely lake in north central Mississippi is the newest and most popular of the Corps of Engineers reservoirs in the Yazoo River Basin. Its waters back up into the Skuna and Yalobusha river valleys, creating a Y-shaped lake with 129 miles of shoreline behind Grenada Dam. There are nearly 10,000 acres of water to explore by boat, and a scenic shoreline of forested hills and rocky slopes to tour by foot or auto. A good highway system crisscrosses the area, and there are numerous hiking trails around the reservoir. The Hugh White State Park, with a lodge and cottages, and Carver Point State Park, border the reservoir. Holly Springs National Forest is not far to the north.

HOW TO GET THERE: From Grenada, take SR 8 east for 3 mi.

FISHING: Bass are the prize catch here—largemouth, spotted, striped, and white. Also plentiful are crappie, bream, perch, and catfish. To give nature a hand, four "fish motels" have been created in the lake by sub-merging old car bodies end on end. Nearly 60 boat launching ramps; rental boats, supplies.

HUNTING: Nearly 75,000 acres of surrounding land make up the Grenada Waterfowl Area. Excellent duck hunting here. Also seasons on deer, turkey, squirrel, and rabbit.

*Unless designated as follows, each project has restrooms, drinking water, and developed campsites. † = no restrooms; ‡ = no drinking water; § = no developed campsites.

CAMPING: Nearly 250 Corps of Engineers campsites, both primitive and developed; facilities include restrooms, showers, drinking water, tables, grills, and launching ramps. Camping also at Hugh White and Carver Point state parks and in Holly Springs National Forest.

OTHER ACTIVITIES: Lake swimming, water skiing, boating; nearly 60 launching ramps, rental boats. Swimming pool at Hugh White State Park. Numerous picnic areas; hiking trails around reservoir.

AREA ATTRACTIONS: Grenada's Historical Museum offers a diverse collection of early Americana and Civil War relics, as well as some rare Indian artifacts. Cottonlandia at Greenwood depicts a pan-orama of Delta life from prehistory to present; it will become a part of the Living Historical Farm now under construction here. Belzoni's Wister Henry Gardens are open to the public year-round.

FOR ADDITIONAL INFORMATION:

Public Affairs Officer
U.S. Army Engineer District, Vicksburg
P.O. Box 60
Vicksburg, MS 39180

Resource Manager
Grenada Lake
P.O. Box 903
Grenada, MS 38901

Forest Supervisor
National Forests in Mississippi
Box 1291
Jackson, MS 39205

OKATIBBEE LAKE

Okatibbee Lake, located on Okatibbee Creek amid the plains of east central Mississippi, is one of the Corps of Engineers' newest projects in the Magnolia State. Its clear waters and irregular, wooded shoreline attract a constantly increasing number of visitors year-round. James Street described this country in his book, Oh, Promised Land, as "warm and lush. And . . . what trees! They just stand there and look way down at you." The roots of the Choctaw Indian nation are deep in this region; Nanih Waiya Mound near Noxapater is the legendary birthplace of the race. Many facilities on 7,000-acre Okatibbee Lake are operated by the Pat Harrison Waterway District, a state agency. Additional opportunities for recreation are provided by Bienville and Tombigbee national forests, each within an hour's drive.

HOW TO GET THERE: From Meridian, take SR 9 north for 7 mi.

FISHING: Bass, bream, and crappie lure anglers to Okatibbee. Nearly 10 launching ramps; lakeside marina, rental boats, supplies.

HUNTING: The State Game and Fish Commission supervises 4,000 acres in the upper reaches of the reservoir as a managed hunting area. Deer, quail, wild turkey, and squirrel roam these woods.

CAMPING: Over 125 campsites, both primitive and developed, are managed by the Pat Harrison Waterway District. Facilities include restrooms, showers, drinking water, tables, grills, playground, dump station, and launching ramps. Additional campsites in Clarkco State Park near Quitman and in the Tombigbee and Bienville national forests.

OTHER ACTIVITIES: Swimming, water skiing, boating; lakeside marina, rental boats. Picnicking and rock and fossil collecting are popular pastimes here. Hiking and horseback riding in nearby national forests. Float trips on the scenic Chickasawhay River.

AREA ATTRACTIONS: The colorful Bound's Water Grist Mill near Vossburg grinds meal on Saturdays. Meridian features a national fish hatchery and Merrehope, one of the town's few remaining antebellum homes. In the same town, the Jimmie Rodgers Memorial honors the "Father of Folk and Country Music"; some of the nation's top country singers gather here for a three-day festival each May. Newton is the home of Mississippi Hydroculture, Inc., a laboratory which grows vegetables in rocks by chemically feeding them; the unique process uses no soil at all.

FOR ADDITIONAL INFORMATION:

Public Affairs Officer
U.S. Army Engineer District, Mobile
P.O. Box 2288
Mobile, AL 36328

Reservoir Manager
Okatibbee Lake
P.O. Box 98
Collinsville, MS 39325

Forest Supervisor
National Forests in Mississippi
Box 1291
Jackson, MS 39205

SARDIS LAKE

An outstanding recreation area in Mississippi's northeastern lakes and hills region, Sardis Lake is fast gaining a reputation as one of the greatest freshwater playgrounds in the Southeast. Vieing for popularity with three other Corps of Engineers reservoirs, all within a 50-mile radius, Sardis attracts well over 2 million visitors annually. Although there are no large cities nearby, visitors will find plenty of modern accommodations and conveniences, including a lodge and cottages at John W. Kyle State Park on the reservoir's north shore. Combine this factor with the tranquility of an 11,000-acre lake surrounded by dense forests, and you have the best of both lives.

HOW TO GET THERE: From Oxford, take SR 6 west about 12 mi., then turn north on SR 315 to dam.

FISHING: Largemouth, spotted, striped, and white bass are plentiful. Anglers will also find bream, crappie, and catfish. Shaded inlets and coves are accessible from shore as well as boat. The Little Tallahatchie River below the dam is also a popular fishing spot. Nearly 50 launching ramps; rental boats, supplies.

HUNTING: Waterfowl hunting at its best in the Sardis Waterfowl Area along the lakeshore; it adjoins the Upper Sardis Wildlife Management Area, where deer, squirrel, rabbit, and quail are abundant. Additional hunting opportunities in a division of Holly Springs National Forest a few miles to the east of the reservoir.

CAMPING: Over 325 Corps of Engineers campsites, both primitive and developed. Facilities include drinking water, restrooms, showers, tables, grills, firewood, playground, and launching ramps. A dump station and more sites are under construction. Group camping area. John W. Kyle State Park on the lakeshore has additional campsites. Private campgrounds in area.

OTHER ACTIVITIES: Lake swimming, pools at the Corps' Rest Haven Campground and at Kyle State Park, water skiing, boating; nearly 50 launching ramps, rental boats. Picnic areas around lakeshore. Nature trails and bike trails on reservoir lands. Hiking in Holly Springs National Forest. Rock and fossil collecting is particularly good in nearby Pontotoc County.

AREA ATTRACTIONS: The town of Holly Springs is an antebellum showplace. Oxford offers more antebellum homes and the University of Mississippi; William Faulkner lived here and used it as the setting for many of his novels. An interesting side trip will take you to Tupelo, headquarters of the Natchez Trace Parkway and birthplace of Elvis Presley.

FOR ADDITIONAL INFORMATION:

Public Affairs Officer
U.S. Army Engineer District, Vicksburg
P.O. Box 60
Vicksburg, MS 39180

Resource Manager
Sardis Lake
Route 2, Box 500
Sardis, MS 38666

ADDITIONAL CORPS OF ENGINEERS LAKES*

Arkabutla Lake and Enid Lake (in NW Mississippi, N of Jackson on I-55)

MISSOURI

LAKE WAPPAPELLO

Along the St. Francis River, near a point where the stream emerges from the Ozark highlands in southeastern Missouri, Lake Wappapello embraces over 180 miles of irregular, wooded shoreline. Not far to the east lies the Mississippi River Valley, which dominates America's midsection. Along part of the reservoir's shoreline are Clark National Forest and Lake Wappapello State Park, creating additional recreational opportunities.

*These projects have restrooms, drinking water, and developed campsites.

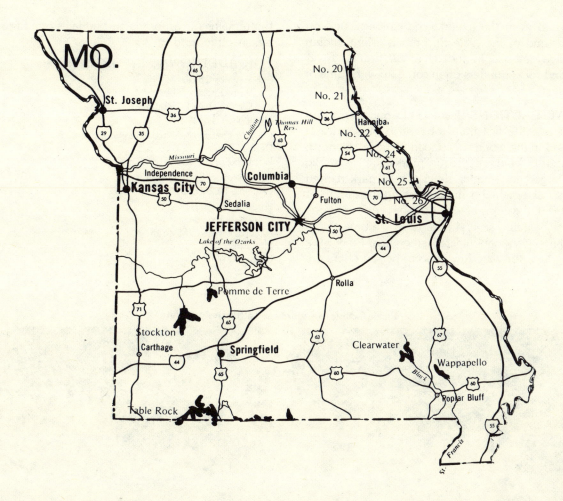

HOW TO GET THERE: From Poplar Bluff, take U.S. 60 west for 1 mi., then north for 8 mi. Continue north for 9 mi. on U.S. 67, then turn east for 8 mi. to the dam.

FISHING: Wappapello has built a reputation as a big bass lake; anglers will find largemouth and white. Crappie, bluegill, and channel catfish also thrive here. Lakeside marina, rental boats, supplies; nearly 20 launching ramps. Float fishing on the Black River in Clark National Forest.

HUNTING: Located on the Mississippi Flyway, Wappapello Lake has long been noted for its fall duck hunting. Hunters will also find deer, wild turkey, and upland game in the Clark National Forest adjoining the reservoir.

CAMPING: Nearly 100 Corps of Engineers campsites, both primitive and developed. Among the facilities are restrooms, showers, drinking water, tables, grills, and launching ramps. Lake Wappapello State Park offers around 125 modern sites, some with electrical hookups. Several campsites in Clark National Forest; some are for boaters on the Black River. Private campgrounds in area.

OTHER ACTIVITIES: Swimming, skin and scuba diving, water skiing, boating; marinas, rental boats, and nearly 20 launching ramps at lakeside. Hiking, particularly in Clark National Forest, and horseback riding. Canoeing and float trips on the Black River in the national forest; canoeing only on the upper reaches of the St. Francis River, which presents a real challenge to even the experienced canoeist, and on the lower end of the Little St. Francis. Many Indian tribes once camped in the fertile St. Francis River Valley, and rockhounders often come up with some of their artifacts.

AREA ATTRACTIONS: Jesse James once terrorized local citizens; the first train robbery in Missouri and the second in the world was staged at Gads Hill north of Piedmont, where a plaque marks the spot. Big Spring, largest in America, is in a state park named after it; it has reached a mind-boggling maximum of 1,130,500,000 gallons of water per day. Adventurous spelunkers will find several caverns in the area to explore. Mingo National Wildlife Refuge near Puxico is a wintering area for waterfowl; also to be seen here are the remains of campsites and artifacts of a historic Indian campground.

FOR ADDITIONAL INFORMATION:

Public Affairs Officer
U.S. Army Engineer District, Memphis
668 Clifford Davis Federal Bldg.
Memphis, TN 38103

Lake Wappapello & Mingo Area Association
Dept. MT-75
Wappapello, MO 63966

Lake Wappapello State Park
Williamsville, MO 63967

Forest Supervisor
Clark National Forest
National Forests in Missouri
P.O. Box 937
Rolla, MO 65401

Picnic facilities are sturdy and attractively modern.

Eerily beautiful at night: Table Rock Lake's dam. (Missouri Conservation Commission/Don Wooldridge photo)

TABLE ROCK LAKE

Majestic bluffs, wooded hills which harbor an abundance of wildlife, and the Ozark Mountains add to the natural beauty of Table Rock Lake itself, a 43,000-acre sprawling giant which follows the valley of the White River in southwestern Missouri. Countless coves, bays, and inlets, several of which reach into Arkansas, are found along an 857-mile shoreline. This lovely reservoir is one of the Corps of Engineers' five most popular, attracting nearly 6 million visitors annually. Table Rock State Park, Mark Twain National Forest, and hundreds of privately owned resorts edge its shores.

HOW TO GET THERE: From Springfield, head south for 45 mi. on U.S. 65.

FISHING: Trout are stocked regularly below the dam, and in the White River tributaries is found some of the best smallmouth bass fishing in the state. Tributary fishing can be combined with a float trip. Lake species include largemouth bass, white bass, crappie, bluegill, and catfish. Nearly 25 launching ramps; lakeside marinas, rental boats, supplies at lakeside.

HUNTING: Rabbits, squirrel, quail, white-tailed deer, and wild turkey test hunters' skills. Mark Twain National Forest borders much of the lake.

CAMPING: Over 1,100 developed Corps of Engineers campsites; facilities include restrooms, drinking water, tables, grills, some showers, electrical hookups, disposal stations, and launching ramps. Table Rock State Park offers around 200 additional modern sites; and there are also some U.S. Forest Service campgrounds along the lakeshore.

OTHER ACTIVITIES: Lake swimming, water skiing, skin and scuba diving, boating of all types; marinas, rental boats, nearly 25 launching ramps. Many scenic picnic areas. Hiking and horseback riding allow visitors to explore the vast shoreline. Drives along forest roads are particularly rewarding during the great splash of color that is autumn in the Ozarks; or take a backpacking trip into the depths of the national forest. The White River tributaries provide some of the best whitewater adventures in the state for canoeists. Whole area good for rockhounding. Nature study is a fascinating hobby; along the Gladetop Trail in the Mark Twain National Forest near Rueter, observant hikers may catch a glimpse of a tarantula or a scorpion.

AREA ATTRACTIONS: A fish hatchery below the dam welcomes visitors. For a glimpse of authentic Ozark mountain characters demonstrating old-time crafts, visit the recreated pioneer village of Silver Dollar City west of Branson. South of the same city is the School of the Ozarks, a liberal arts college primarily run and maintained by students in return for room and board; several buildings are of interest and open to the public. This area was immortalized for all time in the best-selling book, *Shepherd of the Hills,* by Harold Bell Wright; near Branson are the farm where he wrote and an outdoor amphitheater where a dramatization of his story is presented during summer months. One of the famous springs for which Missouri is noted can be seen at Sycamore (near Gainesville); it pours forth 1,200,000 gallons an hour to power old and picturesque Hodgson Mill, which still grinds meal and flour. There are many caves to explore in the area around the reservoir. Springfield, about an hour's drive away, has many attractions. Harry Truman's birthplace at Lamar is open to the public; when Harry was born, his father planted a seedling pine in his dooryard and predicted that both would gain stature in the world.

FOR ADDITIONAL INFORMATION:

Public Affairs Officer
U.S. Army Engineer District, Little Rock
P.O. Box 867
Little Rock, AR 72203

Table Rock Lake Association
P.O. Box 926, Dept. MT-75
Kimberling City, MO 65686

Table Rock State Park
Branson, MO 65616

Forest Supervisor
Mark Twain National Forest
National Forests in Missouri
P.O. Box 937
Rolla, MO 65401

ADDITIONAL CORPS OF ENGINEERS LAKES*

Clearwater Lake (from St. Louis, 120 mi. S via I-55 and U.S. 67 and W on SR 34); Pomme de Terre Lake (from Springfield, 53 mi. N on U.S. 65, 5 mi. W on U.S. 54, 3 mi. S on SR 64); Stockton Lake (from Springfield, 29 mi. N on SR 13, 22 mi. W on SR 32)

*These projects all have restrooms, drinking water, and developed campsites.

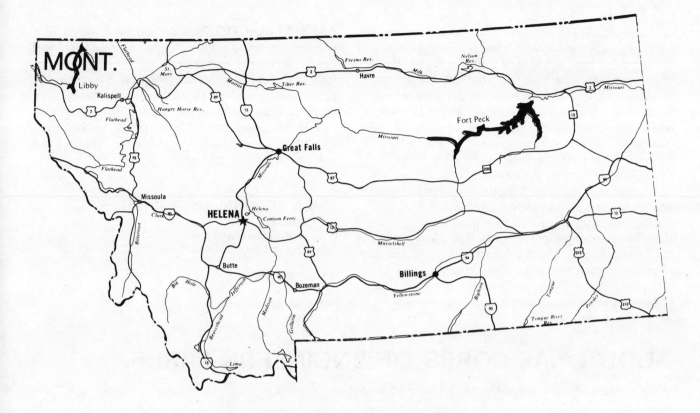

MONTANA

FORT PECK LAKE

Not just the sky, but everything is big in Montana's Missouri River country. Sprawling Fort Peck Lake, with over 1,500 miles of shoreline and 215,000 acres of surface water, is one of the world's largest man-made reservoirs. Along a portion of the south shore are the largest known fossil beds on earth. The vast prairies which surround the lake seem never-ending as they roll away in every direction. Outdoor recreation lovers will find a variety of activities and big country in which to enjoy them.

HOW TO GET THERE: From Glasgow, drive 18 mi. southeast on SR 24.

FISHING: The reservoir and river harbor walleye, northern pike, sauger, rainbow trout, German brown trout, crappie, yellow perch, channel catfish, and the paddlefish. And, like everything else in this state, they run large. Nearly 15 launching ramps; lakeside marina. Float fishing on the Missouri River; ice fishing in winter.

HUNTING: Elk, deer, antelope, turkey, partridge, grouse, pheasant, ducks, and geese provide thrills for hunters.

CAMPING: Over 200 Corps of Engineers campsites, both primitive and developed. Facilities include rest rooms, showers, drinking water, tables, grills, dump stations, some electrical hookups, playground, and launching ramps. Three state parks, as well as private recreation areas, also provide campsites.

OTHER ACTIVITIES: Lake and pool swimming, water skiing, pleasure boating; lakeside marina, nearly 15 launching ramps. Some of the most scenic picnic areas anywhere. There are many backroads here, and the countryside is highly picturesque. Because of large fossil beds, rockhounding is tops. Opportunities for wildlife observation and nature study are enhanced by two areas which border the reservoir—the Charles M. Russell National Wildlife Range and the UL Bend National Wildlife Refuge. A buffalo herd roams a 250-acre range near the dam. Like the reservoir itself, the dam is one of the world's largest, and visitors may tour it. The Corps of Engineers also maintains the Fort Peck Museum at the powerhouse; it displays around 300 specimens of dinosaur bones and other fossils. Horseback riding is an ideal way to see this land. Explore the Missouri River on a float trip or by canoe. In the winter, there are snowshoe hiking, snowmobiling, and cross-country skiing.

AREA ATTRACTIONS: The Fort Peck Indian Reservation, featuring tribal arts and crafts and seasonal events, is near the northeastern shore of the reservoir. Ride a free ferry across the Missouri River near the lake's western end. At Fort Benton, once the head of steamboat navigation on the Missouri River, there's an excellent museum as well as an old steamboat levee which is a National Historic Landmark. Glasgow has a museum and historical courthouse.

FOR ADDITIONAL INFORMATION:

Public Affairs Officer
U.S. Army Engineer District, Omaha
7410 U.S. Post Office and Court House
215 N. 17th St.
Omaha, NE 68102

Reservoir Mgr.
Fort Peck Lake
Fort Peck, MT 59223

ADDITIONAL CORPS OF ENGINEERS LAKES

Libby Dam, or Lake Koocanusa* (from Libby, 17 mi. E on SR 37)

NEBRASKA

HARLAN COUNTY RESERVOIR

In the hilly country of south central Nebraska, Harlan County Lake extends 12 miles up the Republican River Valley. Along the way its waters pierce the mouths of nearly a dozen tributary creeks, creating innumerable quiet coves where one can drop anchor and be insured of some privacy. This is the land which inspired pioneer Dr. Brewster Highley to write "Home on the Range" at his cabin 11 miles south of Franklin. North of the administration area, visitors enjoy watching the goings-on in a busy prairie dog community. Recreation facilities at this large lake which attracts over 1 million visitors annually are operated by the Corps of Engineers.

HOW TO GET THERE: From I-80, between Lexington and Kearney, pick up U.S. 183 south to Alma and the lakeshore.

FISHING: Most common game fish are walleye, white bass, and channel catfish. Largemouth bass, crappie, and northern pike are found in lesser abundance.

---

*This project has restrooms, drinking water, and developed campsites.

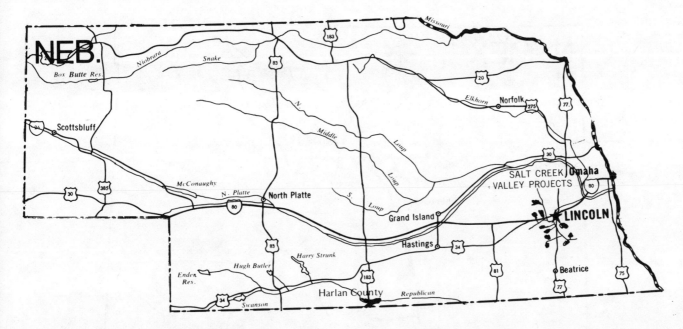

Striped bass have recently been stocked here on an experimental basis. A state record for white bass taken with bow and arrow was set here in the early 1970s. Lakeside marina, rental boats, supplies; nearly half a dozen launching ramps.

HUNTING: Mallard ducks and Canada geese are favorites with waterfowl hunters, and pheasants are the prime upland game. Quail are abundant in certain areas of the project. Other game species include mule deer, cottontail rabbits, and mourning doves. Approximately 16,000 acres of reservoir land are open to public hunting.

CAMPING: Nearly 275 developed Corps of Engineers campsites; facilities include restrooms, drinking water, showers, tables, grills, dump stations, playground, and launching ramps. Alma's City Park offers a limited number of lakeside campsites with water and electrical hookups.

OTHER ACTIVITIES: Swimming, water skiing, boating of all types; lakeside marina, rental boats, nearly half a dozen launching ramps. Picnic areas and hiking trails. Canoeing on both the reservoir and the Republican River.

AREA ATTRACTIONS: At Minden is Harold Warp's outstanding Pioneer Village, where America's progress since 1830 is exhibited in authentic buildings which were moved here from a radius of 350 miles. History passed this way along the banks of the Platte River 50 miles north of Harlan County Reservoir; pioneers heading west on the Oregon Trail along the riverbanks left behind the still-visible rutted tracks of prairie schooners. At the Willa Cather Memorial in Red Cloud, visitors may see the famous author's childhood home and memorabilia.

FOR ADDITIONAL INFORMATION:

Public Affairs Officer
U.S. Army Engineer District, Kansas City
700 Federal Bldg.
601 E. 12th St.
Kansas City, MO 64106

ADDITIONAL CORPS OF ENGINEERS LAKES*

Salt Creek Valley Projects (10 small lakes near Lincoln: Bluestem Lake, near Sprague; Branched Oak Lake, near Raymond; Conestoga Lake§, near Emerald; Holmes Park Lake§, in Lincoln at 70th and Van Dorn Sts.; Olive Creek Lake§, near Kramer; Pawnee Lake, near Emerald; Stagecoach Lake§, near Hickman; Twin Lakes§, near Pleasant Dale; Wagon Train Lake, near Hickman; Yankee Hill Lake‡,§, near Denton)

*Unless designated as follows, each project has restrooms, drinking water, and developed campsites. † = no restrooms; ‡ = no drinking water; § = no developed campsites.

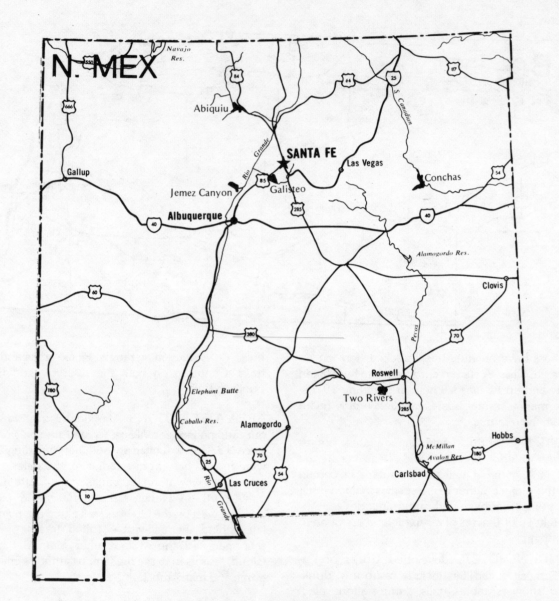

NEW MEXICO

ABIQUIU DAM

High in the northern extremities of New Mexico's Jemez Mountains, in the midst of Carson National Forest's breath-taking beauty, lies Abiquiu Dam. Sitting astride the Rio Chama about 32 miles from its joining with the Rio Grande, Abiquiu Dam does not create a permanent reservoir. However, thanks to a sediment pool maintained on project lands, some water-related activities are possible, and nearly 100,000 people annually come to this reservoir in north central New Mexico to view the scenery and enjoy the recreational facilities.

HOW TO GET THERE: From Espanola, travel 30 mi. north and west on U.S. 84, then 2 mi. south on SR 96.

FISHING: Rainbow and brown trout, bass, and catfish provide the thrills for anglers here. The Corps of Engineers maintains one launching ramp.

HUNTING: No hunting on reservoir lands. Nearby Santa Fe National Forest offers turkey, elk, and deer hunting.

CAMPING: The Corps of Engineers provides just 10 primitive campsites here at present; facilities include drinking water, restrooms, tables, grills, and launching ramps. Several state parks within an hour's drive, as well as the Carson and Santa Fe national forests, offer sites for campers.

OTHER ACTIVITIES: Water skiing and boating; launching ramp. Some swimming is done here, but there's no beach. Several picnic shelters, including some on a rocky rim overlooking the Rio Chama. The Corps of Engineers has constructed a scenic walkway along the bluff rim above the reservoir. Scenic drives throughout area. Full range of winter sports in Carson and Santa Fe national forests. Santa Fe National Forest also offers wilderness pack trips and saddle trails. The whole region is good for fossil collecting; some have been found which date back 200 million years.

AREA ATTRACTIONS: In the high green valleys of Carson National Forest are villages where only Span-ish is spoken, while the Santa Fe National Forest contains living Indian pueblos. Just north of Abiquiu, off U.S. 84, is the Ghost Ranch, a living museum exhibiting animals and plants indigenous to the land. It completely surrounds 1.25-acre Beaver National Forest, which demonstrates the multiple-use features of national forests; life-size figures add realism to scenes of various recreational pursuits. The Cumbres & Toltec Scenic Railroad, steam-powered and narrow-gauge, takes its passengers on a 64-mile trip through some of the most beautiful and remote mountain scenery in the country; it operates during summer months from Chama, NM, to Antonito, CO. Nowhere else in North America did prehistoric man develop living quarters like the unique man-made "cave rooms" in the cliffs of Frijoles Canyon; they can be seen today at the Bandelier National Monument near Los Alamos. A unique bit of modern history was made in Los Alamos itself; the first atomic bomb was built here. Tourists may learn more about it at the city's Scientific Laboratory Museum.

FOR ADDITIONAL INFORMATION:

Public Affairs Officer
U.S. Army Engineer District, Albuquerque
P.O. Box 1580
Albuquerque, NM 87103

Regional Forester
National Forests in New Mexico
Federal Bldg., Sixth Floor
517 Gold Ave., S.W.
Albuquerque, NM 87101

CONCHAS LAKE

Situated in the South Canadian River canyon of northeastern New Mexico, amid dramatic desert surroundings, Conchas Lake provides a 15-mile-long water pathway which has become one of the state's most popular recreational playgrounds. Artifacts of the earliest known people in the Western Hemisphere have been found in this part of New Mexico. In more recent times, the famed Santa Fe Trail passed through this area. Recreation facilities at Conchas Lake lie within the three units of Conchas Lake State Park. They include a lodge, fisherman cabins, restaurants, and a mile-long paved airstrip.

HOW TO GET THERE: From Tucumcari, take SR 104 north and west for 31 mi.

FISHING: Conchas is regarded as one of the state's best fishing lakes. Fishing for largemouth bass, bluegill, crappie, channel catfish, and walleye is rated excellent. Over half a dozen launching ramps; lakeside marinas, rental boats, supplies.

HUNTING: Along the rocky canyons of the Canadian River is a herd of wild Barbary sheep, imported from North Africa; and there is a special limited season for

Water skiing is a popular activity at many Corps of Engineers lakes. (Corps of Engineers photo)

bagging one. Hunters will also find antelopes, deer, quail, rabbit, and migratory waterfowl in this area.

CAMPING: Over 150 campsites in Conchas Lake State Park, both primitive and modern. Facilities include restrooms, drinking water, tables, grills, full hookups, dump station, and launching ramps.

OTHER ACTIVITIES: Swimming at three beaches and a swimming pool, skin diving, water skiing, and boating; lakeside marina, rental boats. Canoes, houseboats, and sailboats are seen alongside small outboards and large cabin cruisers. Picnic areas, hiking trails, and a nine-hole golf course are available for landlubbers, as well as playgrounds for children. Winter sports are also popular here.

AREA ATTRACTIONS: The Fort Union National Monument northeast of Las Vegas lay along the Old Santa Fe Trail; many of its adobe walls are still standing, and visitors may learn about its place in history within a museum at the visitor center. Las Vegas itself was once the gathering place for the scum of the Southwest; many of them were hanged from a windmill that used to be on the plaza, a spot which is visited by tourists today. Tucumcari and Las Vegas both have historical museums. East of Santa Rosa are a U.S. fish hatchery and Blue Hole, a miniature spring-fed lake stocked with colorful fish; both may be viewed by the public. The Kiowa National Grasslands border the Canadian River southeast of Springer.

FOR ADDITIONAL INFORMATION:
Public Affairs Officer
U.S. Army Engineer District, Albuquerque
P.O. Box 1580
Albuquerque, NM 87103

ADDITIONAL CORPS OF ENGINEERS LAKES*

Galisteo Dam (from Santa Fe, 23 mi. S on U.S. 85, 5 mi. E on project access road); Jemez Canyon Dam (from Bernalillo, 2 mi. W on SR 44, 6 mi. N on project access road); Two Rivers Dam (from Roswell, 14 mi. W on U.S. 70/380, 7 mi. S on project access road). Reservoirs at all these projects are dry, except for limited storage during spring runoffs.

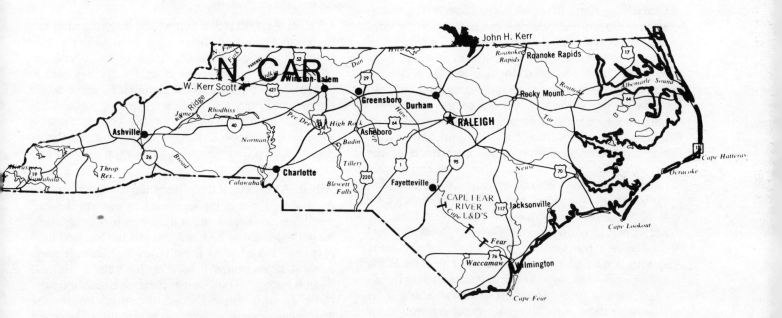

NORTH CAROLINA

JOHN H. KERR RESERVOIR

A sprawling giant shared by North Carolina and Virginia, the John H. Kerr Reservoir was once known, and is often still referred to, as Buggs Island Lake. Its 800 miles of shoreline encompass a lake with nearly 50,000 acres of surface area, creating a 39-mile-long playground on the Roanoke River for over 3 million annual visitors. Located in the gently rolling hill country of the northern portion of the Piedmont Plateau, Kerr Reservoir lies in a peaceful valley characterized by alternating woods and meadows. The highly cultured Occoneechee Indian tribe dominated this

*These projects have restrooms but no drinking water or developed campsites.

region until around 300 years ago, and many Indian artifacts are still found in the area. Some are on display in the resource manager's office near John H. Kerr Dam. Numerous towns and villages in both North Carolina and Virginia offer every service and accommodation any visitor could desire. Occoneechee and Staunton River state parks, both in Virginia, have been established at the water's edge. Though neither have lodges, Staunton River offers several lakeside vacation cabins. An excellent highway system crisscrosses the adjacent region, making the lake easily accessible from any direction.

HOW TO GET THERE: From Henderson, NC, take I-85 north. The reservoir lies less than 5 mi. west of this highway, and there are numerous turnoffs to various access points.

FISHING: Kerr Reservoir has built a reputation around its record striped bass catches. When the Roanoke River was sealed off by Kerr Dam, a number of striped bass that had moved upstream to spawn were trapped. They've adapted very successfully, and today it's not uncommon to see a school of several hundred come to the surface in pursuit of threadfin shad. Largemouth bass and crappie fishing are rated excellent, and there's some fair panfish angling as well. Nearly a dozen launching ramps maintained by the Corps of Engineers; several other state and private ramps. Lakeside marinas, rental boats, supplies. Because of unannounced and swift changes in water level, the areas just above and below Kerr Dam are not open to fishermen. Gaston Reservoir just downstream offers some good fishing for the same species you'll find in Kerr, as well as for pickerel. In the spring, spawning striped bass in excess of 10 pounds head up the Roanoke and Dan rivers above Kerr Reservoir; these same rivers are good year-round for smallmouth and rock bass and are stocked with trout in portions of their upper reaches.

HUNTING: In designated areas around reservoir, including the wooded areas of Occoneechee State Park, VA, and at numerous North Carolina game lands in Granville, Vance, and Warren counties south of the reservoir. The white-tailed deer is the number one big game quarry. Also some wild turkey. Small game includes dove, grouse, rabbit, quail, squirrel, fox, and migrating waterfowl.

CAMPING: Over 300 Corps of Engineers campsites, both primitive and developed. Facilities include restrooms, drinking water, showers, tables, grills, dump stations, and launching ramps. Two state parks on the Virginia shore, Staunton River and Occoneechee, have campgrounds, and the latter has a few sites with water and electricity. The Kerr Reservoir Development Commission, an official state agency, has developed several recreation areas with campsites along the North Carolina lakeshore. Many private campgrounds on the lake and nearby.

OTHER ACTIVITIES: Swimming in lake and in the Staunton River State Park pool (VA); water skiing, boating; lakeside marinas, rental boats. Some organized boat rides. Weekend sailing regattas held at various times throughout the summer. Nearly a dozen Corps of Engineers launching ramps; others are operated by the states and private concessionaires. Picnicking, horseback riding, and hiking along shoreline. Staunton River State Park has a tennis court and children's playgrounds.

AREA ATTRACTIONS: (Virginia) Danville, one of the nation's major tobacco auction centers, has 17 warehouses where visitors may watch the fall sales. Near Brookneal is Red Hill, the last home and burial place of Patrick Henry. (North Carolina) Raleigh offers three state museums (one each for history, art, and natural history). The Nuclear Reactor Building on the North Carolina State University campus houses the first college-owned reactor, open without restriction to the public; it's devoted exclusively to peacetime development of the atom. Durham nearby is another huge tobacco auction center with September through December sales open to the public; year-round tours of a cigarette factory are available at Liggett & Myers.

FOR ADDITIONAL INFORMATION:
Public Affairs Officer
U.S. Army Engineer District, Wilmington
P.O. Box 1890
Wilmington, NC 28401
Resource Manager
John H. Kerr Reservoir
Rte. 1, Box 76
Boydton, VA 23917
Kerr Reservoir Development Commission
Rte. 3, Box KRDC
Henderson, NC 27536

W. KERR SCOTT RESERVOIR

In North Carolina's lovely Yadkin River Valley between the towns of Wilkesboro and Ferguson, W. Kerr Scott Reservoir twists its way upstream for nearly seven miles through the forested foothills of the Blue Ridge Mountains. With its 55-mile shoreline and conservation pool of 1,470 acres, the reservoir is one of the Corps of Engineers' smaller projects, but it lies in a scenic area which offers both an abundance of recreational opportunities and a wide range of facilities. A restaurant, lodge, and vacation cottages are planned for construction soon near the lake. It was in the Yadkin River Valley that Daniel Boone met and married his Rebecca. Both were Wilkes County residents at the time, and they settled for a while near the mouth of Beaver Creek a few miles upstream from the site of W. Kerr Scott Dam.

HOW TO GET THERE: From Wilkesboro, head west 5 mi. on SR 268.

FISHING: The largemouth bass is king here, but crappie and panfish are also available. There are excellent public trout streams in the mountains to the west of the reservoir. Nearly 10 launching ramps; lakeside marina, rental boats, supplies.

HUNTING: White-tailed deer are the most sought-after big game. Hunters will also find wild turkey, rabbit, squirrel, grouse, and raccoon on reservoir hunting lands and on the Thurmond Chatham State Game Land in Wilkes County north of the reservoir.

CAMPING: Around 150 Corps of Engineers campsites, both primitive and developed. Facilities include restrooms, drinking water, showers, heated dressing rooms, tables, grills, dump stations, and launching ramps. Marley's Ford State Park southwest of Wilkesboro offers a limited number of primitive sites. Private campgrounds in area.

OTHER ACTIVITIES: Swimming, water skiing, boating; lakeside marina, rental boats, nearly 10 launching ramps. Canoeing on the Yadkin River. Picnicking along the lakeshore. Hiking in an area rich in natural beauty. There's gold in the hills of western North Carolina—also rubies and emeralds—and visitors are welcome to look for them, as well as other interesting rocks and minerals.

AREA ATTRACTIONS: The Corps of Engineers Visitor Center at W. Kerr Scott Dam features historical exhibits pertaining to the local area; included are portraits, dioramas, statues, and a representation of the syllabary Sequoya developed for the Cherokees. Moravian and Cascade falls are nearby. The Blue Ridge Parkway, one of the country's outstanding scenic drives, winds through the Blue Ridge Mountains and portions of the Pisgah National Forest to the west. Along the parkway, which is administered by the National Park Service, and in the national forest are recreation areas which offer horseback riding, mountain climbing, hiking, and skiing in the winter. Linville Gorge Wilderness, Linville Falls, and Linville Caverns are spectacular scenic attractions in the national forest. Mt. Mitchell, also in the forest, is the highest point east of the Mississippi (6,684 ft.). At Grandfather Mountain near Linville, visitors with strong constitutions may walk over a deep gorge by way of a mile-high swinging bridge. If you like a hint of the mysterious, visit Brown Mountain off SR 181 between Morganton and Lenoir on most any clear night, and you may be able to see the strange, recurring lights which have been attributed by some to UFOs.

FOR ADDITIONAL INFORMATION:

Public Affairs Officer
U.S. Army Engineer District, Charleston
P.O. Box 919
Charleston, SC 29402

Reservoir Manager
W. Kerr Scott Reservoir
P.O. Box 182
Wilkesboro, NC 28697

ADDITIONAL CORPS OF ENGINEERS LAKES*

Cape Fear River Locks and Dams§: Pool No. 1 (from Wilmington, 2 mi. W on U.S. 74-17, 16 mi. W on U.S. 74-76, then about 15 mi. NW on SR 87, 1 mi. NE on state road to lock); Pool No. 2 (from Elizabethtown, about 1 mi. SE on SR 87, 1 mi. NE on state road to lock); William O. Huske Pool (from Fayetteville, about 17 mi. S on SR 87)

*Unless designated as follows, each project has restrooms, drinking water, and developed campsites. † = no restrooms; ‡ = no drinking water; § = no developed campsites.

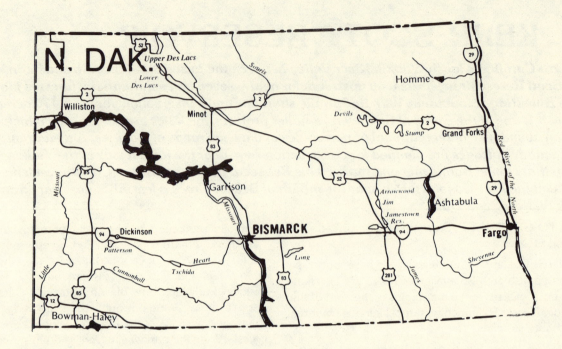

NORTH DAKOTA

GARRISON DAM

Named in honor of the remarkable Shoshone Indian girl who helped lead the Lewis and Clark Northwest Expedition from these plains to the Pacific Coast, Lake Sakakawea is a giant of a reservoir formed by Garrison Dam along the Missouri River Valley of northwestern North Dakota. Eons of history have left their mark here; sites of early Indian culture and of old trading and Army posts are scattered throughout the area. Theodore Roosevelt loved this wild country and once made it his home. Today, the two units of Theodore Roosevelt National Memorial Park lie in the heart of the savagely beautiful North Dakota Badlands to the south and west of Lake Sakakawea. Recreation facilities on the lakeshore are primarily operated by the Corps of Engineers, the state of North Dakota, local governments, and an Indian tribal council.

HOW TO GET THERE: From Bismarck, take U.S. 83 north for 70 mi., then head west on SR 200 for 10 mi. to Garrison Dam.

FISHING: Lake Sakakawea is one of the state's top angling waters. Walleye and sauger fishing is outstanding, with several record sauger taken here. Northern pike also provide excellent sport. Tailwaters below the dam produce walleye, northern pike, channel catfish, and rainbow trout, while the main stream of the Missouri River is good for channel cats, northerns, and white bass. Nearly 20 launching ramps; lakeside marina. Ice fishing in winter.

HUNTING: Due to a large concentration of national wildlife refuges and waterfowl rest areas, no better

There is plenty of room for family fun at lakeside picnic areas. (Corps of Engineers photo)

waterfowl hunting is found anywhere in the country. Other available game includes white-tailed and mule deer, sharptail grouse, Hungarian partridge, pheasant, fox and gray squirrel, cottontail rabbit, and coyote.

CAMPING: Over 250 Corps of Engineers campsites, both primitive and developed; facilities include restrooms, showers, drinking water, tables, grills, some electrical hookups, dump stations, and launching ramps. Over 100 more campsites in several state-operated recreation areas. Four Bears Park, which also offers camping, is managed by the Indian tribal council on the Fort Berthold Indian Reservation. Several local governments also operate campgrounds on the lakeshore.

OTHER ACTIVITIES: Swimming, water skiing, boating; lakeside marina, nearly 20 launching ramps. Canoeing on Little Missouri River. Excellent rockhounding area. Horseback riding is very popular here; rental horses and organized trail rides nearby. Lots of picnic areas. There's a scenic drive through the North Unit of Theodore Roosevelt Memorial Park, as well as an extensive system of backcountry trails. Another scenic drive skirts Killdeer Mountain near the town of Killdeer. Wildlife which may be seen by visitors includes beaver, porcupines, prairie dogs, bobcats, golden eagle, and western diamondback rattlers. There's a small herd of wild horses in Roosevelt Park, as well as wild buffalo, antelope, and bighorn sheep. Interesting rock formations and petrified wood abound in region. Ice skating in winter.

AREA ATTRACTIONS: At New Town is an all-Indian museum owned and maintained by area tribes; visitors may see and purchase crafts work. Lake Ilo and Audubon national wildlife refuges are nearby. A federal fish hatchery below the dam welcomes visitors. Near Beulah, the public may watch lignite coal being mined by the open-pit method, and near Stanton they may tour the world's largest lignite-burning electric plant.

FOR ADDITIONAL INFORMATION:
Public Affairs Officer
U.S. Army Engineer District, Omaha
7410 U.S. Post Office and Court House
215 N. 17th St.
Omaha, NE 68102
Area Engineer
Lake Sakakawea—Garrison Dam
Riverdale, ND 58565

LAKE ASHTABULA

Amid the slightly rolling prairie of east central North Dakota, Lake Ashtabula follows the winding valley of the Sheyenne River for 27 miles. The wooded valley and blue waters create a refreshing oasis in the plains, providing one of the few contradictions to a countryside dominated by wheat fields and a vast expanse of sky. Recreation facilities are either operated by the Corps of Engineers or locally managed. This lovely, scenic lake is rarely crowded, being overshadowed in popularity by its giant counterparts along the Missouri River to the west.

HOW TO GET THERE: From Valley City, take River Rd. for 11 mi. northwest.

FISHING: Very good in recent years for northern pike and walleye, the state's two major game fish. Other species are crappie, yellow perch, white bass, and bullhead. Nearly 10 launching ramps; lakeside marina, rental boats, supplies. Devoted fishermen go at it through the ice in winter.

HUNTING: Hunters will find some top waterfowl gunning here. White-tailed deer, fox and gray squirrel, cottontail rabbit, and coyote are also hunted.

CAMPING: Nearly 75 Corps of Engineers developed campsites; facilities include restrooms, showers, drinking water, tables, grills, and launching ramps. Locally-managed campsites also number around 75 and provide similar facilities.

OTHER ACTIVITIES: Swimming, water skiing, boating; lakeside marina, rental boats. Canoeing on the Sheyenne River. Also picnicking, hiking, and rockhounding. A scenic drive follows SR 1 south of Valley City. Ice skating and snowmobiling in winter.

AREA ATTRACTIONS: A federal fish hatchery below the dam is open to the public. The Northern Prairie Wildlife Research Center is a major research facility operated by the U.S. Fish and Wildlife Service and related to the preservation of wildlife resources; guided tours are available for the public. The world's largest buffalo, also near Jamestown, is a three-story-high statue of an American bison that will leave children gasping. Bonanzaville at West Fargo is an outstanding living museum which recalls the famous nineteenth-century Bonanza Farm era in North Dakota; it includes a pioneer village and the Regional Museum of the Red River and Northern Plains. Near Carrington is the Arrowhead National Wildlife Refuge.

FOR ADDITIONAL INFORMATION:

Public Affairs Officer
U.S. Army Engineer District, St. Paul
1135 U.S. Post Office and Custom House
St. Paul, MN 55101

Reservoir Manager
Lake Ashtabula
Valley City, ND 58072

ADDITIONAL CORPS OF ENGINEERS LAKES*

Bowman-Haley Lake (from Bowman, 13 mi. S on U.S. 85, 9 mi. E and 2 mi. S on gravel county road); Homme Lake (from Grand Forks, 39 mi. N on U.S. 81, 24 mi. W on SR 17)

OHIO

ATWOOD LAKE

Deep in the wooded hills of east central Ohio lies Atwood Lake, one of a series of Corps of Engineers lakes in Muskingham River country. Backing up Indian Fork of the Conotton Creek, the reservoir offers over 1,500 acres of water playground surrounded by a scenic, forested shoreline. Luxurious Atwood Lake Lodge is the focal point of a recreation park which has become one of Ohio's most popular. This area, along with nine other reservoirs, is primarily administered by the Muskingham Conservancy District, a public corporation organized under Ohio law, with state and federal cooperation. Extensive recreation facilities attract visitors both winter and summer.

HOW TO GET THERE: From Dover, head 7 mi. north on SR 800, then 9 mi. east on SR 212.

FISHING: Angling for largemouth bass, northern pike, catfish, and panfish is generally good. Two lakeside marinas, rental boats, supplies; three launching ramps. State fishing laws apply. Ice fishing in winter.

HUNTING: Deer, rabbit, squirrel, grouse, pheasant, and waterfowl around the lake and in the dense forests nearby. Hunting is regulated by the Ohio Division of Wildlife.

CAMPING: Nearly 600 developed campsites are managed by the Muskingham Conservancy District. Facilities include drinking water, restrooms, showers, firewood, playground, dump stations, and boat launching ramps. Most sites have electricity; many edge the reservoir. Vacation cabins are also available. Several private campgrounds in vicinity.

OTHER ACTIVITIES: Winter or summer, there's plenty to do here. Swimming in the lake, water skiing, boating. Rent a canoe, houseboat, sailboat, pontoon

---

*These projects all have restrooms, drinking water, and developed campsites.

boat, or fishing boat at the lake's marinas. Atwood Lake is noted for exceptionally good sailing March through November. During summer months visitors may sign up for scenic cruises or enjoy nature programs presented in the park amphitheater. There are public golf and tennis facilities at Atwood Lodge. Explore the surrounding country by foot, bicycle, horseback, or auto, and enjoy a picnic lunch with a lake view. Ice fishing and ice boat racing grow more popular each year. A ski slope, ice skating rink, and toboggan and sled runs are near the lodge. No snow? They'll make some for you.

AREA ATTRACTIONS: Two state memorials in the area are restorations of long-dissolved religious communities; Zoar Village is near the town of the same name, and Schoenbrunn Village is just outside New Philadelphia. Another state memorial with the lyrical name of Gnadenhutten commemorates the site of a third historic village near Urichsville; a museum is open to the public. History is presented from another aspect at Warther's Museum in Dover; most of the exhibits are the work of one man who was billed during his lifetime as the world's master carver, and few will dispute this claim after a visit here. The National Pro Football Hall of Fame and President McKinley's grave and memorial are at Canton. Near Wilmot are the Alpine-Alpha, a swiss cheese factory which offers self-guided tours, and the Stark Wilderness Center for nature education. Lt. Col. George A. Custer, noted Indian fighter, was born in New Rumley; a monument and exhibit pavilion mark his birthplace. The Swiss-Amish village of Sugarcreek exudes old European rural charm, complete with horsedrawn buggies.

FOR ADDITIONAL INFORMATION:

Public Affairs Officer
U.S. Army Engineer District, Huntington
P.O. Box 2127
Huntington, WV 25721

Manager, Land Department
Muskingham Conservancy District
New Philadelphia, OH 44663

Atwood Lake Lodge
P.O. Box 96
Dellroy, OH 44620

BERLIN LAKE

Berlin Lake, because of its clean water, scenic setting and easy accessibility to several large cities, attracts over 1.5 million visitors annually to enjoy a full range of recreational activities. An impoundment of the Mahoning River in historic northeastern Ohio, Berlin Lake is the largest man-made reservoir entirely within this corner of Ohio. Though surrounded by populous towns which lie within an hour's drive in every direction, the lake maintains an air of serenity. There are few developments along its shore, enhancing an outdoors experience here.

HOW TO GET THERE: From Deerfield, take SR 224 east for 2 mi.

FISHING: Largemouth bass, white bass, crappie, walleye, and channel cat await anglers. Some freshwater drum in the Mahoning River. Many coves and inlets offer quiet hideaways for serious fishermen. Lakeside marina, rental boats, supplies; almost 10 launching ramps. Ice fishing in winter.

HUNTING: In designated areas around reservoir for waterfowl, deer, grouse, quail, rabbit, and squirrel. A field trial area has been established by the Ohio Division of Wildlife.

CAMPING: Nearly 500 Corps of Engineers campsites, both primitive and developed. Facilities include drinking water, restrooms, tables, grills, playground, dump station, and launching ramps. West Branch State Park to the northwest has additional family campsites and some primitive walk-in sites. Several private campgrounds near lake.

OTHER ACTIVITIES: Swimming, water skiing, boating; lakeside marina, rental boats, almost 10 launching ramps. Shaded picnic areas. Ice skating when weather conditions permit. Hiking on the Buckeye Trail; this footpath, which spans the state, skirts the eastern side of the reservoir.

AREA ATTRACTIONS: Stan Hywett Hall and Gardens, a 65-room mansion filled with priceless antiques and works of art, is in Akron. The city is the world's leading producer of rubber products, and you can tour the major plants. The Goodyear World of Rubber, opposite company headquarters, has a display of space-age products. Also near Akron is the American Indian Art Hall of Fame. Of special interest at Youngstown's Butler Institute of American Art are the miniatures of American presidents. Mill Creek Park in the same city offers scenic drives and foot trails, a formal garden, and a nature education center. President McKinley's tomb in Canton is a state memorial; the National Pro Football Hall of Fame is in the same city. Sea World in Aurora is the home of Shamu, the killer whale, and a great white shark. The renowned Kenley Players draw visitors from across the country to Packard Music Hall in Warren to view Broadway plays and musicals.

FOR ADDITIONAL INFORMATION:
Public Affairs Officer
U.S. Army Engineer District, Pittsburgh
Federal Bldg., 1000 Liberty Ave.
Pittsburgh, PA 15222
Reservoir Manager
Berlin Lake
Deerfield, OH 44411

DEER CREEK LAKE

Lush, verdant woodlands edge much of the 20-mile shoreline of Deer Creek Lake, which lies along a Scioto River tributary in the rural flatlands of south central Ohio. The region is steeped in Indian history. Tecumseh, the great Shawnee chief, and Chief Logan of the Mingos lived near here with their tribes, and the Shawnee-Miami Trail passed through the Deer Creek project area. There are several prehistoric Indian sites on project lands. A later event in history, the "Teapot Dome" scandal, is believed to have been hatched in an old lodge across the lake from the dam; it was once the private property of President Warren G. Harding and is currently undergoing restoration.

Visitors should take a look at the dam which impounds Deer Creek; it received an Award of Merit for Engineering Design in 1970. Another attraction on reservoir lands is a unique research project instituted by the Corps of Engineers. An "on-land (spray irrigation) wastewater disposal system" will provide data for future sewage treatment facilities elsewhere in the United States.

HOW TO GET THERE: From Circleville, head west on U.S. 22 for about 13 mi.; turn north on SR 207 to project.

FISHING: Deer Creek Lake is rated good for bass, bluegill, and crappie. A weir dam constructed as part of the downstream fishing provides excellent stream fishing in Deer Creek below the dam. Two launching ramps, rental boats, supplies. Ice fishing in winter.

HUNTING: Wildlife areas at the extreme north and south ends of the reservoir are state-managed for public hunting. Sportsmen gun for deer, rabbit, squirrel, grouse, and waterfowl.

CAMPING: Over 230 family campsites and a group camping area in Deer Creek State Park. Facilities include restrooms, showers, drinking water, waste-water drains, tables, fire rings, and dump stations. Most have electricity at the site. One shower building has been designed to accommodate campers in wheelchairs. Several private campgrounds in surrounding area.

OTHER ACTIVITIES: Swimming, water skiing, boating; rental boats, two launching ramps. Sailing and canoeing are popular. Over a mile of picnic area extends along the south shore, and other picnic areas are scattered around the lake. Ice skating in winter.

AREA ATTRACTIONS: Four scenic loop drives traverse Paint Valley to the summit of Ohio's most spectacular hills near Bainbridge. Seven Caves nearby offer hiking and spelunking. Chillicothe features *Tecumseh,* a stunning outdoor drama, and Mound City Group National Monument, the largest known concentration of Hopewell Indian burial mounds in the country. Fairfield County ranks first in Ohio and second in the nation with its total of 37 covered bridges. A few miles south of Circleville, you can watch the pressing of wine and taste the final product at the Shawnee Vineyards. Though seasonal, Circleville's four-day Pumpkin Show each October deserves mention as one of the U.S.A.'s greatest festivals.

FOR ADDITIONAL INFORMATION:

Public Affairs Officer
U.S. Army Engineer District, Huntington
P.O. Box 2127
Huntington, WV 25721

Reservoir Manager
Corps of Engineers
Scioto Area
P.O. Box 638
Circleville, OH 43113

WEST FORK OF MILL CREEK LAKE

A small, serene lake surrounded by low, wooded hills and a well-manicured landscape, 188-acre Mill Creek Lake is a part of Winton Woods Park about 15 miles north of downtown Cincinnati in southeastern Ohio. Recreation facilities are administered by the Hamilton County Park District under a license issued by the government. The park itself completely surrounds the village of Greenhills, one-time home of the late writer and television personality, Rod Serling. Over a million visitors use the wide range of facilities annually.

HOW TO GET THERE: From Cincinnati, take I-75 north 20 mi. to I-275; turn west for 5 mi. to Winton Rd. exit, and turn south for another 2.5 mi. to the park. Though a roundabout route, it's the fastest way to get there.

FISHING: Fishing is not the main attraction here, as it is at many Corps of Engineers reservoirs, but there are some crappie, bluegill, and bass. Except for senior citizens 65 and over, who may fish from the bank in designated areas, boat fishing only is permitted. Some fishing ponds have been established especially for the handicapped. Motors limited to 3.6 HP or less; you may rent one or bring your own, but boats must be rented. Small fee for adult fishing permits; children free in designated areas. Ice fishing in winter.

HUNTING: No hunting.

CAMPING: Over 50 family sites, both primitive and developed, and one group camping area. Facilities include restrooms, showers, drinking water, playground, tables, grills, firewood, disposal station.

OTHER ACTIVITIES: Rental boats only (rowboats, motorboats, pedal boats); rent or bring your own motor (3.6 HP limit). Sightseeing tours aboard 100-passenger paddlewheel boats. Family picnic areas have grills; groups may reserve enclosed fee lodges or shelters in advance. Planned recreation available. A complete horseback riding center, 18-hole golf course, baseball diamond, model plane flying area, archery range, and outdoor education area. Also horseshoes, hiking, rockhounding; ice skating and snow skiing when weather permits. Outdoor movies and band concerts during summer. Canoeing on the Little Miami River.

AREA ATTRACTIONS: Cincinnati has a myriad of attractions. Sports fans may visit Riverfront Stadium, home of professional baseball and football in the Queen City. Mt. Adams is an interesting hilltop area reminiscent of San Francisco. Cross the Ohio River on Anderson Ferry, about six miles west of town. The city's zoo, zoo opera, art museum, and city parks are all outstanding. Fort Ancient State Memorial near Lebanon and Miamisburg Mound State Memorial at the city of the same name feature Indian mounds and relics. Dayton's U.S. Air Force Museum is a delight for visitors of any age. Youngsters in the eight-to-eighty category will enjoy the vast King's Island Family Amusement Park off I-71 near Lebanon.

FOR ADDITIONAL INFORMATION:
Public Affairs Officer
U.S. Army Engineer District, Louisville
P.O. Box 59
Louisville, KY 40201
Hamilton County Park District
10245 Winton Rd.
Cincinnati, OH 45231

ADDITIONAL CORPS OF ENGINEERS LAKES*

Beach City Lake§ (from New Philadelphia, 12 mi. N on SR 250); Bolivar Lake§ (from Bolivar, 11 mi. S on I-77, 1 mi. SE on SR 212 and County Rd. 103); Charles Mill Lake (from Lucas, 5 mi. E on U.S. 30, 4 mi. S on SR 603); Clendening Lake (from Urichsville, 15 mi. S on SR 800); Delaware Lake (from Columbus, 35 mi. N on U.S. 23); Dillon Lake (from Zanesville, 4 mi. NW on SR 146); Dover Dam§ (from Dover, 4 mi. NE on SR 800); Leesville Lake (from New Philadelphia, 13 mi. E on SR 39, 4 mi. SE on SR 212); Michael J. Kirwan Reservoir (from Warren, 13 mi. W on SR 5, follow signs to dam); Mohawk Lake (from Coshocton, 13 mi. W on U.S. 36, 2 mi. NW on SR 715); Mosquito Creek Lake§ (from Cortland, 2 mi. W on SR 305); Piedmont Lake (from Cambridge, 24 mi. NE on U.S. 22); Pleasant Hill Lake (from U.S. 30, 6 mi. S on SR 60, SW on SR 95); Senecaville Lake (from Cambridge, 6 mi. S on I-77, 6 mi. E on SR 313); Tappan Lake (from Urichsville, 6 mi. E on U.S. 250); Tom Jenkins Dam, or Burr Oak Lake (from Athens, 20 mi. N on U.S. 33 and SR 13); Wills Creek Lake (from Coshocton, 7 mi. S on SR 76, 2 mi. SW on County Rd. 497)

---

*Unless designated as follows, each project has restrooms, drinking water, and developed campsites. † = no restrooms; ‡ = no drinking water; § = no developed campsites.

OKLAHOMA

BROKEN BOW LAKE

Lovely Broken Bow Lake, embraced by 180 miles of pine-studded shoreline, lies in the heart of southeastern Oklahoma's magnificent scenery. Fed by the clear Mountain Fork River, which flows from the foothills of the Ouachita Mountains, and dotted by a chain of islands which once were mountaintops along the riverbank, Broken Bow is bounded at its northern end by the primitive beauty of the McCurtain County Wilderness Area. French and Spanish explorers who passed through this country in centuries past and the Choctaw Indians who once owned this land saw scenery much like that seen by today's visitors. Beavers Bend, along Broken Bow's shoreline, is one of the finest parks in the entire state system.

HOW TO GET THERE: From Idabel, head north for 21 mi. on U.S. 259 and 2 mi. east on SR 295A.

FISHING: Linson Creek, a Broken Bow tributary, boasts some of the largest sunfish in the state. Other species inhabiting the lake are largemouth, small-mouth, and spotted bass, crappie, and catfish. Nearly 10 launching ramps; lakeside marina, rental boats, supplies.

HUNTING: A State Game Management Area which adjoins the lake is managed primarily for deer, elk, and turkey. Other species available here and on designated reservoir lands are rabbit, squirrel, quail, ducks, geese, and dove.

CAMPING: Around 50 Corps of Engineers campsites, both developed and primitive. Facilities include drinking water, restrooms, showers, tables, grills, and launching ramps. At Beavers Bend State Park, approximately 90 modern campsites offer some electrical hookups and a dump station. Private campgrounds in area, including a KOA with swimming pool and full hookups.

OTHER ACTIVITIES: Swimming, water skiing and boating; lakeside marina, rental boats. Mountain Fork River is one of the state's best canoeing streams. Rockhounders will find agates in the Kiamichi Mountains which border the reservoir on the north. Picnic areas are complemented by magnificent views. Hikers will find easy nature trails; one features the largest white oak tree in Oklahoma, the other a beaver dam and lodge. Backpacking is permitted in the McCurtain Wilderness Area, but be on the lookout for resident rattlesnakes, copperheads, and cottonmouths. Guided tours into McCurtain are available on weekends. Since almost every kind of timber to be found in Oklahoma is found on reservoir lands, and since wildlife species are varied and abundant, this is a good spot for nature study. One of mid-America's loveliest scenic highways, the Talimena Skyline Drive, follows SR 1 along the crest of Winding Stair Mountains east of Talihina.

AREA ATTRACTIONS: Oklahoma's oldest tree, a huge cypress that took root before the birth of Christ, is east of the town of Broken Bow on U.S. 70. A drive through the portion of Ouachita National Forest north of the reservoir leads to Heavener and Runestone State Park, where tourists may see stones bearing characters reputedly inscribed by Vikings when they visited Oklahoma almost 1,000 years ago. Near Poteau is the world's highest hill, Cavanal, which misses by one foot the 2,000-foot height necessary for qualification as a mountain. There's a marvelous view from its crest.

FOR ADDITIONAL INFORMATION:
Public Affairs Officer
U.S. Army Engineer District, Tulsa
P.O. Box 61
Tulsa, OK 74102

CANTON RESERVOIR

Nestled amid the high plains of northwestern Oklahoma, with a tree-studded shoreline which slopes gently to the water's edge, 8,000-acre Canton Reservoir extends up the valley of the North Canadian River for approximately 6 miles. It takes its name from a pioneer Army post, the "Cantonment of the Canadian River," which once stood near the west shore of the lake; today the decaying remains of a single structure mark the site. The garrison was manned from 1879 to 1882, when it became a Mennonite school for Cheyenne Indian children. The Cheyenne Dictionary, considered one of the greatest and most scholarly works about American Indian languages, was written by a Mennonite missionary during his tenure here. Canton Reservoir today is one of the Sooner State's most popular outdoor recreation spots, attracting around 1.5 million visitors annually. For those who don't care to camp, Roman Nose State Park a few miles to the southeast offers a resort lodge. Visitors will also find housekeeping cottages at the state park, as well as at Sportsman's Park, one of the recreation areas operated by the Corps of Engineers at Canton Reservoir. There's a paved and lighted airstrip near the dam.

HOW TO GET THERE: From Fairview, head south for 13 mi. on SR 58, then 2 mi. west on SR 58A.

FISHING: Canton is noted for its walleye fishing, and every spring local merchants sponsor a walleye rodeo. Largemouth and white bass, crappie and channel catfish also provide angling thrills. About a dozen launching ramps; rental boats, supplies.

HUNTING: A 17,000-acre public hunting area adjoins the lake. Sportsmen will find bobwhite quail, deer, waterfowl, squirrel, wild turkey, and dove.

CAMPING: About 160 developed Corps of Engineers campsites; facilities include restrooms, showers, drinking water, tables, grills, dump stations, electrical hookups, playgrounds, and launching ramps. Roman

Nose State Park, not far to the southeast, has additional sites with full hookups.

OTHER ACTIVITIES: Lake swimming, water skiing, boating; rental boats, about a dozen launching ramps. Visitors will find a softball field and tennis courts near the dam. Roman Nose State Park a few miles away offers a swimming pool, horseback riding, and hayrides.

AREA ATTRACTIONS: At Roman Nose State Park visitors will see typical western landscape with exposed white strata of gypsum, which is abundant in the area; those who care to sample its peculiar taste may try the gypsum-flavored water here. Just north of the park, at Southard, a large plant converts the gypsum into sheetrock. Near Orienta are the Glass Mountains, a photographer's delight; these bright red buttes, interlaced with layers of white gypsum, have eroded away over the centuries to reveal millions of selenite crystals which reflect sunlight like glass mirrors. Near Waynoka, children from eight to eighty enjoy playing in the huge sand dunes along the Cimarron River at the Little Sahara Recreation Area; visitors may also ride dune buggies and view the park's animals, which include Indian and longhorn cattle, camels, and goats. There's a well-preserved sod house, once used by a pioneer homesteader, near Cleo Springs. Travelers who happen to be in the area during early April might enjoy watching the goings-on at Okeene's Rattlesnake Roundup; hunters from all over the country capture hundreds of rattlers as they're emerging from hibernation in the rocky escarpments near the town.

FOR ADDITIONAL INFORMATION:
Public Affairs Officer
U.S. Army Engineer District, Tulsa
P.O. Box 61
Tulsa, OK 74102

Resident Engineer
Western Resident Office
Corps of Engineers
P.O. Box 67
Canton, OK 73724

GREAT SALT PLAINS RESERVOIR

Great Salt Plains Lake, on the Salt Fork of the Arkansas River in northwestern Oklahoma, lies in a unique region which played a valuable role in the history of the West. Once the great animals of the Plains migrated here because of the salt supply, creating a rich hunting ground which was the cause of many Indian battles. Later the salt flats proved just as valuable to Kansas and Texas cattlemen. Today the area is popular with rockhounders, who avidly seek the selenite crystals which thrive in certain soil concentrations. Crystal digging is allowed on Saturdays, Sundays, and holidays from April 1 to October 15 each year under rules established by the Great Salt Plains National Wildlife Refuge, which administers most of the reservoir lands here. Recreational facilities, including housekeeping cabins, are within Great Salt Plains State Park and are located near the dam.

HOW TO GET THERE: From Enid, drive 36 mi. north, then west on U.S. 64 to Jet; dam is 8 mi. north of here on SR 38.

FISHING: Despite the saline content of the water, channel catfish abound in the lake. Striped bass, walleye, and blue catfish have also been stocked in the reservoir. Trotline fishing is profitable at times. Four launching ramps provide access to this 9,300-acre lake; rental boats and supplies.

HUNTING: Famed for its goose hunting, this reservoir attracts nearly 40,000 geese annually. No hunting on government land here, except for occasional special seasons set up on the wildlife refuge, but nearby landowners lease blinds. Deer, quail, wild turkey, and numerous species of small animals and birds are abundant.

CAMPING: Nearly 200 campsites, both primitive and developed. Facilities include restrooms, showers, drinking water, tables, grills, playgrounds, some electrical hookups, and launching ramps. Campgrounds are operated by the state of Oklahoma and the Great Salt Plains National Wildlife Refuge.

OTHER ACTIVITIES: Lake swimming, water skiing, boating (rental boats available). Picnic areas near dam. Great rockhounding; in addition to the selenite crystals on reservoir land, the area around Waynoka produces a unique specimen known as the rattlesnake egg. The Great Salt Plains National Wildlife Refuge, one of a chain of refuges along the Continental Central Flyway, hosts a multitude of migratory waterfowl and has a resident population as well. It is one of the nation's most popular wintering areas for golden and bald eagles, and more than 250 other species have been observed here. Hikers enjoy the refuge's Eagle Roost Nature Trail, where they may see beaver, white-tailed deer, wild turkey, egrets, and herons. An observation tower near the refuge entrance overlooks the level plains, covered by a thin layer of salt for as far as the eye can see.

AREA ATTRACTIONS: Blackwell's several dog kennels are training sites for many of the greyhounds seen on American racing tracks. At Waynoka's Little Sahara Recreation Area, children of all ages may play on the huge sand dunes which border the Cimarron River and view camels, goats, and Indian and longhorn cattle. Northeast of Cleo Springs is a preserved sod house built by one of the area's original homesteaders. South of Freedom is Alabaster Caverns State Park, where visitors may tour the world's largest known alabaster cave, complete with sleeping bats, and view the rugged beauty of Cedar Canyon. Oklahoma's Glass Mountains, near Orienta, are of particular interest to photographers; these bright red buttes have eroded over the centuries to reveal millions of selenite crystals which reflect sunlight like glass mirrors.

FOR ADDITIONAL INFORMATION:

Public Affairs Officer
U.S. Army Engineer District, Tulsa
P.O. Box 61
Tulsa, OK 74102

Great Salt Plains National Wildlife Refuge
Route 1, Box 49
Jet, OK 74739

KEYSTONE LAKE

Amid a setting of wooded shoreline, high bluffs, and rolling hills, the blue-green waters of Keystone Lake form a twisting path along the Arkansas River channel in northeastern Oklahoma. An irregular, 330-mile shoreline, pierced by scores of coves and inlets, encircles the 26,300-acre lake. The Osage and Creek Indians contributed to the rich history of this region, as did Washington Irving, noted American author, who wrote of his 1832 western adventures here in A Tour of the Prairies. *A myriad of excellent recreational facilities, managed by the Corps of Engineers, the state of Oklahoma, and the cities of Cleveland and Sand Springs, await today's visitors.*

HOW TO GET THERE: From Tulsa, travel west for 19 mi. on U.S. 64 (Keystone Expressway).

FISHING: Fishermen flock to Keystone Lake for the striped bass, which fisheries biologists believe to exceed 40 pounds. White and black bass, crappie, and catfish are also abundant. Nearly 30 launching ramps; lakeside marina, rental boats, supplies.

HUNTING: Principal game species include white-tailed deer, bobwhite quail, mourning dove, ducks, geese, cottontail rabbit, and squirrel. Approximately 17,000 acres of reservoir land are open for public hunting.

CAMPING: Over 300 Corps of Engineers campsites, both primitive and developed; facilities include rest-rooms, drinking water, showers, tables, grills, dump stations, some electrical outlets, and launching ramps. Additional sites available in Keystone State Park, Walnut Creek State Park, and Feyodi Creek State Recreation Area. Several private campgrounds nearby.

OTHER ACTIVITIES: Swimming, water skiing, boating; lakeside marina, rental boats. Sailing is extremely popular here. Many scenic picnic areas. The Washington Irving Scenic Nature Trail passes over varied terrain along the lakeshore; beaver cuttings may be observed enroute.

AREA ATTRACTIONS: The Pawnee Bill Museum and State Park near Pawnee honor a showman and partner of Buffalo Bill; buffalo and longhorn cattle may be seen in the park. Pawhuska is the capital of the Osage nation, one of the wealthiest Indian tribes in America; around 250,000 head of beef cattle range the vast grasslands outside of town. The works of noted artists of the American West are exhibited at Tulsa's Gilcrease Institute of American History and Art. Also in Tulsa is Mohawk Park, one of the country's largest municipally owned parks, which contains a zoo, golf course, bridle trails, and a small buffalo herd. The Sun Oil Company and the Texaco refinery, both in Tulsa, offer free tours.

FOR ADDITIONAL INFORMATION:
Public Affairs Officer
U.S. Army Engineer District, Tulsa
P.O. Box 61
Tulsa, OK 74102

Keystone Resident Office
U.S. Army, Corps of Engineers
Route 1
Sand Springs, OK 74063

LAKE EUFAULA

Among the oak-clad hills of south central Oklahoma, giant Lake Eufaula reaches into numerous draws and crevices to create 600 miles of shoreline along the valleys of three major streams. The Canadian, North Canadian, and Deep Fork rivers all flow into the reservoir known in the Sooner State as the "Gentle Giant." The powerful Creek and Choctaw Indian nations once lived here; the famous outlaw gangs of Belle Starr and the James Brothers came here to hide out; Civil War battles were fought on this soil. Today's visitors will find over twenty Corps of Engineers public-use areas and two major state parks, with luxury lodges and cottages, offering a vast variety of recreational opportunities.

HOW TO GET THERE: From McAlester, go 23 mi. north on U.S. 69, 16 mi. east on SR 9, and 6 mi. north on SR 71 to dam.

FISHING: A variety of species, including largemouth, smallmouth, spotted, striped, and white bass; channel, blue, and flathead catfish; and black and white crappie, call Lake Eufaula home. Several heated docks insure comfortable winter fishing. Nearly 40 launching ramps; lakeside marinas, rental boats, supplies.

HUNTING: On reservoir lands open for public hunting, hunters will find bobwhite quail, deer, cottontail rabbits, squirrels, ducks, geese, and mourning dove.

CAMPING: Nearly 500 developed Corps of Engineers campsites; facilities include restrooms, showers, drinking water, tables, grills, and launching ramps. Almost 175 additional campsites, some with full hookups and dump stations, are found in Arrowhead and Fountainhead state parks, while numerous private campgrounds offer yet more sites with a choice of conveniences.

OTHER ACTIVITIES: Swimming, water skiing, boating of all types; lakeside marinas, rental boats. Both state parks have swimming pools. Cruise boat rides available. Canoes take to the reservoir and its tributaries. Picnic areas everywhere. Four major hiking trails around reservoir; one of them, near the south end of the dam, is a prime source for arrowhead hunters. Other Indian artifacts are also found in the area. Horseback riding, 18-hole golf, tennis, archery, bicycling, and game courts are available at state parks. Winter sports have come into their own here; sledding and ice skating in sheltered coves. Many scenic drives in area; Talimena Skyline Drive, a national scenic highway built just for the view, heads east from Talihina, OK, on the crest of Winding Stair Mountains.

AREA ATTRACTIONS: Infamous outlaw Belle Starr is buried one mile east of Eufaula Dam. Creek Nation Crafts, Inc., at Checotah displays and sells authentic Indian pottery. The Creek Nation Council House and Museum at Okmulgee is a registered National Historic Landmark. Autumn visitors here may enjoy a unique rodeo held over Labor Day weekend near McAlester; it takes place within the walls of Oklahoma State Penitentiary, and the participants are all inmates.

FOR ADDITIONAL INFORMATION:
Public Affairs Officer
U.S. Army Engineer District, Tulsa
P.O. Box 61
Tulsa, OK 74102

Lake Eufaula Association
P.O. Box 822
Eufaula, OK 74432

A quiet, scenic boat-ride on one of the many man-made lakes.

OOLOGAH LAKE

Lying in the lush and beautiful Verdigris River Valley of northeastern Oklahoma, Oologah Lake is encompassed by 75 miles of rambling, tree-dotted hills. Once this region was dominated by the Cherokee Indian nation, but the rich heritage which survives it is perhaps overshadowed by the legacy of a single man. This was the land which shaped and molded humorist Will Rogers, who was born in a ranch house which would now be at the bottom of Oologah Lake if it had not been relocated on a nearby hilltop site. Rogers often said that, though he was born halfway between Claremore and Oologah, "I usually say I was born in Claremore for convenience, because nobody but an Indian can pronounce Oologah." The state of Oklahoma maintains his birthplace within the boundaries of Will Rogers State Park adjacent to the reservoir.

HOW TO GET THERE: From Tulsa, head 30 mi. north on U.S. 169, then 3 mi. east on SR 88.

FISHING: Oologah is a top bass lake, thanks to the habitat produced by flooded brush and timber along its shoreline. Anglers are also challenged by crappie and channel and flathead catfish. Half a dozen launching ramps; lakeside marina.

HUNTING: The Oologah Game Management Area is managed primarily for game such as white-tailed deer, fox and gray squirrel, cottontail and swamp rabbit, bobwhite quail, and raccoon. Other areas nearby are also managed for migratory birds, such as ducks, geese, and mourning dove.

CAMPING: Over 225 Corps of Engineers campsites; facilities include restrooms, drinking water, showers, tables, grills, dump stations, and launching ramps. More campsites available in Will Rogers State Park and at private campgrounds in the nearby area.

OTHER ACTIVITIES: Swimming, water skiing, boating; lakeside marina. Spacious picnic area. The Skull Hollow Nature Trail leads through woods and meadows, past a rocky gorge and a boulder-strewn bluff. Many opportunities for nature study.

AREA ATTRACTIONS: Claremore is noted for several artesian mineral wells with properties beneficial in the treatment of certain ailments. Will Rogers' plans to build a home near here were thwarted by his untimely death; his tomb and a sprawling memorial now occupy the site. In the heart of Claremore is the J. M. Davis Gun Collection, one of the world's largest private collections. The Tom Mix Museum in Dewey displays memorabilia of the famous silent-movie star and a former marshal of this town. Bartlesville has several attractions of interest: Price Tower downtown was designed by Frank Lloyd Wright; Oklahoma's first commercially important oil well has been reproduced as a memorial to oil men; Woolaroc Ranch offers a fascinating museum which traces the advance of civilization in western America, as well as 3,400 acres of rugged timberland roamed by many varieties of wild animals. Several oil companies in the vicinity offer free tours. Tulsa, with its diverse attractions, has something to offer to everyone.

FOR ADDITIONAL INFORMATION:
Public Affairs Officer
U.S. Army Engineer District, Tulsa
P.O. Box 61
Tulsa, OK 74102

Oologah Lake Resident Office
P.O. Box 38
Oologah, OK 74053

TENKILLER FERRY LAKE

A beautiful reservoir which winds along the twisting valley of the spring-fed Illinois River for more than 25 miles, Tenkiller Ferry Lake lies in the heart of the rugged Cookson Hills of northeastern Oklahoma. Some 130 years ago, the Cherokee Indians came west to this land from Georgia to build their nation anew. Here in the Ozark foothills they invented a syllabary, produced

Oklahoma's first newspaper in both Cherokee and English, and developed a tribal government and schools. Though much of this region has changed since then, the cliffs, forests, and meadows along Tenkiller's 130-mile shoreline remain very much as they were when the Cherokees first saw them over a century ago. Recreational facilities, however, offer the ultimate in modern convenience.

HOW TO GET THERE: From Muskogee, travel 21 mi. southeast on SR 10, then 7 mi. east on SR 10A.

FISHING: Largemouth and smallmouth bass, white bass, crappie, catfish, bream, and walleye inhabit the reservoir, while rainbow trout thrive in the tailwaters of the dam. Float trips draw many fishermen to the upper Illinois River in the Tahlequah area. Heated fishing docks insure comfortable winter fishing. About 35 launching ramps on reservoir; lakeside marinas, rental boats, supplies.

HUNTING: High success rates for deer are recorded on public lands around the reservoir and in nearby public hunting areas. Small game available includes squirrel; cottontail, swamp, and jackrabbits, and migratory game birds such as dove, quail, woodcock, duck, and geese.

CAMPING: Over 600 developed Corps of Engineers campsites; facilities include restrooms, showers, drinking water, tables, grills, electrical hookups, and launching ramps. Tenkiller State Park offers over 300 more campsites, some with full hookups and dump stations; and there are private campgrounds as well near the lakeshore.

OTHER ACTIVITIES: Swimming, water skiing, boating of all types; lakeside marinas, rental boats. Clear waters invite skin and scuba divers. Tenkiller State Park has a swimming pool. The Illinois River is a famous canoeing and float stream, with campsites along the way. Hiking and picnicking are great family activities.

AREA ATTRACTIONS: At Tahlequah, former capital city of the Cherokee Indian nation, tribal government buildings built in the last half of the nineteenth century still stand; they include the Supreme Court Building, the Capitol Building, and the National Prison. Tsa-La-Gi nearby is a living replica of an early Cherokee Indian village; the outdoor theater here offers a look at the tribe's history in a summer drama entitled *The Trail of Tears.* The Cherokee Indian Weavers are east of Tahlequah on U.S. 62. A National Military Cemetery at Fort Gibson contains a monument to renowned Seminole warrior Billy Bowlegs and the burial place of Sam Houston's Cherokee wife. The home of Sequoya, northeast of Sallisaw, is a national shrine to the inventor of the Cherokee alphabet.

FOR ADDITIONAL INFORMATION:

Public Affairs Officer
U.S. Army Engineer District, Tulsa
P.O. Box 61
Tulsa, OK 74102

Lake Tenkiller Association
P.O. Box Ten K
Cookson, OK 74427

ADDITIONAL CORPS OF ENGINEERS LAKES*

Chouteau Lock and Dam (from Muskogee, 7 mi. N on U.S. 69, 3 mi. SE on access road); Fort Gibson Lake (from Muskogee, 9 mi. N on SR 80); Fort Supply Lake (from Woodward, 13 mi. NW on U.S. 183 and 270, and SR 3); Heyburn Lake (from Tulsa, 25 mi. W on U.S. 66, 7 mi. W on access road); Hugo Lake (from Hugo, E on U.S. 70, about 8 mi. to dam); Hulah Lake (from Bartlesville, 12 mi. N on U.S. 75, 12 mi. W on SR 10); Kaw Lake (from

*Unless designated as follows, each project has restrooms, drinking water, and developed campsites. † = no restrooms; ‡ = no drinking water; § = no developed campsites.

Some of the picnic areas and boat ramps at Robert S. Kerr Lake.

Pine Creek Lake retains a wilderness flavor.

Ponca City, about 8 mi. E on U.S. 60); Newt Graham Lock and Dam‡, § (from Tulsa, 25 mi. E on SR 33, 7 mi. S on county road); Pine Creek Lake (from Idabel, 18 mi. W on U.S. 70 to Valliant, 2 mi. N on SR 98, 7 mi. N on county road); Robert S. Kerr Reservoir (from Fort Smith, AR, 22 mi. W on I-40, 8 mi. S on U.S. 59); W. D. Mayo Lock and Dam§ (from Fort Smith, AR, 12 mi. W on SR 9, 4 mi. N on county road); Webbers Falls Lock and Dam (from Muskogee, 20 mi. S on Muskogee Turnpike, 5 mi. E on U.S. 64, 2 mi. N on SR 10); Wister Lake (from Wister, 2 mi. E on U.S. 270)

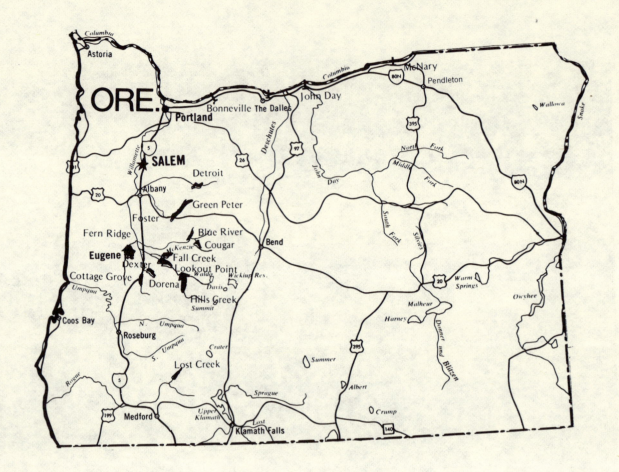

OREGON

JOHN DAY LOCK AND DAM (LAKE UMATILLA)

Behind John Day Dam, Lake Umatilla extends for miles through the scenic Columbia River Valley, offering 240 miles of shoreline for outdoor fun. Visitors come from all over to watch the passing parade of commercial traffic, to take a look at migrating fish through below-water-level windows, and to sample Corps of Engineers recreation areas. Migrating waterfowl sometimes darken the sky as they travel the Pacific Flyway, and Umatilla National Wildlife Refuge has been established on both the Washington and Oregon shores to provide a resting place for them.

HOW TO GET THERE: From The Dalles, OR, head east on I-80 for 22 mi.

FISHING: The Columbia River is famous for its salmon fishing, during the spring and fall runs; the three species found here are chinook, coho, and an occasional sockeye. Another quarry is the fighting steelhead trout. Sturgeon and shad are also available. Lakeside marina; almost 15 launching ramps.

HUNTING: Many upland game birds are taken from the steep canyons and brushy areas along the Columbia River and its estuaries. Waterfowl species found here include mallards, Canada geese, and wood ducks. Hunting in season on Umatilla National Wildlife Refuge.

CAMPING: Nearly 200 Corps of Engineers campsites, both primitive and developed. Facilities include restrooms, showers, drinking water, tables, grills, and launching ramps. One Corps of Engineers campground is accessible by water only. Private campgrounds in area.

OTHER ACTIVITIES: Swimming, water skiing, boating; lakeside marina, almost 15 launching ramps. Some good sailing stretches. Canoeing on John Day River, a scenic waterway. Boaters enjoy the unique experience of locking through John Day Dam, one of the largest power producers in the world. Several picnic areas. Wildlife observation and nature study at Umatilla National Wildlife Refuge. Saddle and pack trips, as well as snow skiing, in Umatilla National Forest nearby.

AREA ATTRACTIONS: (Oregon) The Umatilla Indian Reservation is near Pendleton. Pendleton is also the headquarters for the Umatilla National Forest, which features dude ranches, hot sulphur springs, and the scenic Kendall-Skyline Forest Road along the summit of the Blue Mountains. The Peace Pipe Museum on the Umatilla County Fairgrounds at Hermiston is one of the most dramatically displayed historical collections in the West. (Washington) About eight miles west of John Day Dam, on SR 14, is America's Stonehenge at Maryhill; a replica of the mysterious 4,000-year-old English Stonehenge, it was erected as a World War I memorial. The outstanding Maryhill Museum of Fine Arts is three miles west of this memorial.

FOR ADDITIONAL INFORMATION:
Public Affairs Officer
U.S. Army Engineer District, Portland
P.O. Box 2946
Portland, OR 97208

LAKE BONNEVILLE

A bonanza of natural beauty and outstanding recreation opportunities await visitors to 48-mile-long Lake Bonneville. This major tourist mecca, impounded behind Bonneville Lock and Dam, lies in the breathtakingly beautiful Columbia River Gorge about 50 miles east of Portland, OR. Sheer rock walls rise to 2,000 feet around the lake, while in the distance snow-covered Mt. Hood dominates the Oregon horizon, a landmark of enduring majesty. No less than eight state parks border the shores of Lake Bonneville, one in Washington and seven in Oregon.

HOW TO GET THERE: From Portland, 35 mi. east on I-80N.

FISHING: Columbia River salmon are nationally famous; anglers will find three species here—chinook, coho, and sockeye (though the latter are few and far between). Also available are steelhead trout, shad, and sturgeon. Lakeside marina; nearly half a dozen launching ramps.

HUNTING: No hunting on project lands. In surrounding areas, white-tailed deer, pheasant, quail, and grouse are plentiful. Conboy Lake National Wildlife Refuge near Glenwood, WA, offers hunting for upland game, big game, and waterfowl.

CAMPING: Nearly 175 campsites, both primitive and developed, provided in Oregon and Washington state parks along reservoir. Facilities include restrooms, showers, tables, grills, drinking water, playground, some utility hookups and launching ramps. Additional campsites in Washington's Pinchot National Forest and Oregon's Mount Hood National Forest, as well as several private campgrounds in area.

OTHER ACTIVITIES: Swimming, water skiing, boating; lakeside marina, several launching ramps. Numerous hiking trails and scenic drives. Picnic spots aplenty; tables near the dam permit visitors to eat while watching river traffic pass through the huge locks. Bonneville Dam at the lake's western end features ponds in a landscaped garden setting and an underwater window for fish viewing. At the western end of the reservoir is The Dalles Dam, where the Corps of Engineers provides free train rides and guided tours to visitor areas. Sightseers will see some scenery for which superlatives are not superlative enough—deep gorges, towering rock formations, and lovely waterfalls (Multnomah Falls, second highest in the United States, is the granddaddy of them all). Several fish hatcheries, all of which welcome the public, are on the lakeshore. At Cascade Locks Marine Park in Oregon, stop and see the locomotive that pulled ships through river

Phillipi Park on the John Day River is accessible only by boat.

locks over 100 years ago. Washington's Pinchot National Forest and Oregon's Mount Hood National Forest offer saddle and pack trips, a full array of winter sports, mountain climbing and huckleberry picking in late summer, as well as many natural phenomena to see and tour. The Pinchot National Forest offers something else—a Pacific Northwest legend that refuses to die. The vast woods are believed by many to be a home for Bigfoot, a giant, apelike creature who acts more human than simian. Portions of the Pacific Crest Trail pass through both forests.

AREA ATTRACTIONS: Portland's small but unique People's Park in the heart of the city shouldn't be

missed. The city also has many gardens which burst into bloom during summer months. Warm Springs Indian Reservation in Oregon and Yakima Indian Reservation in Washington allow visitors. Against a hillside near the confluence of the Columbia and Little White Salmon rivers in Washington is the longest active lumber flume in the United States. Fort Vancouver National Historic Site lies in the heart of the present-day city of Vancouver, WA.

FOR ADDITIONAL INFORMATION:
Public Affairs Officer
U.S. Army Engineer District, Portland
P.O. Box 2946
Portland, OR 97208

ADDITIONAL CORPS OF ENGINEERS LAKES*

Blue River Lake§ (from Springfield, 40 mi. E on SR 126); Cottage Grove Lake (from Cottage Grove, 6 mi. S on I-5 and county road); Cougar Lake§ (from Springfield, 50 mi. E on SR 126); The Dalles Lock and Dam, or Lake Celilo (from The Dalles, 3 mi. E on I-80N); Detroit Lake (from Salem, 46 mi. E on SR 22); Dexter Lake (from Eugene, 10 mi. SE on SR 58); Dorena Lake (from Cottage Grove, 8 mi. E on county road); Fall Creek Lake§ (from Eugene, 20 mi. E on SR 58); Fern Ridge Lake§ (from Eugene, 14 mi. W on SR 126); Foster Lake (from Sweet Home, 4 mi. NE on SR 20); Green Peter Lake (from Sweet Home, 10 mi. NE on SR 20 and county road); Hills Creek Lake (from Eugene, 43 mi. SE on SR 58); Lookout Point Lake (from Eugene, 12 mi. SE on SR 58); McNary Lock and Dam, or Lake Wallulua§ (from Umatilla, 2 mi. E on SR 730)

PENNSYLVANIA

ALLEGHENY RESERVOIR (KINZUA DAM)

Twelve thousand acres of clear water, rimmed by steep banks and forest-draped mountains, spiral out of the northeast corner of Allegheny National Forest in Pennsylvania into New York State. Indians, missionaries, and explorers passed this way long ago when this land was a primeval wilderness; soldiers, lumbermen, and oil bargemen wrote later pages in the region's history. Though the original forest has dwindled in size, the woods remaining around much of the lake's shoreline are lovely and unspoiled. Much of the reservoir in New York State borders the Allegany

*Unless designated as follows, each project has restrooms, drinking water, and developed campsites. † = no restrooms; ‡ = no drinking water; § = no developed campsites.

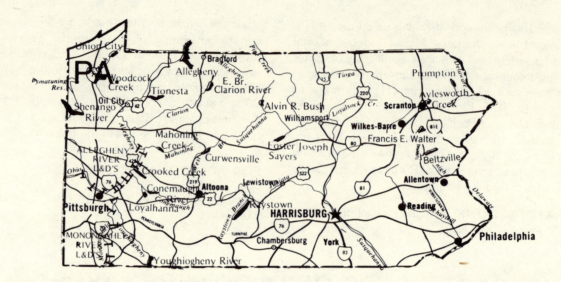

(yes, it's spelled differently in the Empire State) Indian Reservation of the Seneca Nation, and the tribe is responsible for the development of reservation lands.

Recreation facilities in Pennsylvania have been constructed at a number of sites on the reservoir by the Corps of Engineers and the U.S. Forest Service, while Allegany State Park adjoins the lake in New York on state-owned land.

HOW TO GET THERE: From Warren, PA, go 9 mi. east on SR 59.

FISHING: Species found here include trout, smallmouth bass, walleye, northern pike, muskellunge, crappie, and yellow perch. Shore fishing is limited due to inaccessibility, but there is excellent bank fishing at Devil's Elbow a few miles above the dam. Purchase a license from the state in which you fish or from the Seneca nation's office near Salamanca, NY, if you choose to wet your line in reservation waters. Nearly half a dozen launching ramps around reservoir; lakeside marina, rental boats and supplies. Ice fishing in winter.

HUNTING: Both large and small game thrive in the varied habitats here. Among them are white-tailed deer, rabbit, fox, raccoon, porcupine, squirrel, and woodchuck. Game birds include turkey, pheasant, quail, grouse, woodcock and migratory waterfowl.

CAMPING: The Corps of Engineers has no campsites here, but there's no shortage. There are over 400 Forest Service campsites on the Pennsylvania section of the reservoir, with more sites at Allegany State Park and the Seneca nation's Quaker Bridge Area, both in New York. In addition, the area includes many private campgrounds. Visitors will find most any type of facilities they desire. Several Forest Service campgrounds are accessible only to boat campers and, in some instances, backpackers.

OTHER ACTIVITIES: Nearly half a dozen swimming beaches; Scandia Mountain Recreation Area along the west bank of the Allegheny Reservoir has a heated swimming pool. Water skiing and boating; lakeside marina, rental boats and supplies. Scenic areas for picnicking. Hiking possibilities are virtually limitless; ten miles of the North Country Trail in the national forest are intended as a segment of our country's proposed national trail system. While hiking, be on the lookout for the scores of wildlife species which inhabit this area; wild bear roam the woods, and there's a heron rookery within the Tracy Ridge Wilderness in the national forest. Horseback riders will find stables, trails, and organized trail rides. Many good canoeing streams. The Allegheny Reservoir Scenic Drive encircles the lake, and an airport at Scandia offers flights for sightseers. Winter sports include skiing (downhill and cross-country), snowmobiling, and snowshoeing.

AREA ATTRACTIONS: In Kinzua Bridge State Park near Mt. Jewett, PA, is a historic bridge which is the second highest in the United States. The grave of Chief Cornplanter, a friend of George Washington, is on the north side of Willow Bay on the reservoir. A new fish hatchery complex has been constructed by the U.S. Bureau of Sport Fisheries and Wildlife just downstream from the dam.

FOR ADDITIONAL INFORMATION:

Public Affairs Officer
U.S. Army Engineer District, Pittsburgh
Federal Bldg., 1000 Liberty Ave.
Pittsburgh, PA 15222

Forest Supervisor
Allegheny National Forest
P.O. Box 847
Warren, PA 16365

Allegany State Park Commission
Salamanca, NY 14779

Kinzua Dam Vacation Bureau
Box 844
Warren, PA 16365

BELTZVILLE LAKE

Embraced by the towering wooded hills along Pohopoco Creek in the Lehigh River Basin of east central Pennsylvania, Beltzville Lake lies at the edge of the famous Pocono Mountains year-round resort playground. Shallow, semitropical seas covered this area approximately 350 million years ago, forming coral reefs similar to those existing today in the Florida Keys. The visitor center displays specimens of the types of fossils to be found in this area, and a spot has been set aside at a Corps of Engineers recreation area where they may be collected by the public. An authentic covered bridge serves as a pedestrian crossing between the Pine Run Cove bathing beach and picnic grove; dating back to 1841, it originally spanned Pohopoco Creek in an area now inundated by reservoir waters. The Corps of Engineers operates day-use areas only at present, with more facilities planned for the future. Beltzville State Park also edges this 947-acre reservoir.

HOW TO GET THERE: From Lehighton, take U.S. 209 east for 4 mi.; turn north on dam road.

FISHING: Excellent crappie waters here. Also good for yellow perch, largemouth bass, channel catfish, walleye, pickerel, tiger muskie, and alewives. Trout is stocked below the dam. Two launching ramps. Ice fishing in winter.

HUNTING: Land has been set aside here for wildlife habitat. The hunter will find ducks, geese, pheasant, doves, deer, and rabbit.

CAMPING: No camping facilities at present, but development is still underway and the Corps of Engineers plans campsites for the future. There are many private campgrounds in the surrounding area, including a KOA at the town of Jim Thorpe.

OTHER ACTIVITIES: Swimming on sand beaches, water skiing, boating; two launching ramps. Picnicking in scenic, wooded areas. Hiking along the lake's borders; one trail passes an old slate quarry. A portion of the Appalachian Trial passes through the Pocono Mountains east of here. Fossil collecting near Pine Run Cove. The Lehigh River west of the reservoir is a whitewater paradise for experienced canoeists only; kayaks or closed-deck canoes a must. All types of winter sports in surrounding area.

AREA ATTRACTIONS: The Pocono Mountains are hard to surpass simply for the scenery alone, but there are other attractions as well. A good place to begin a

Most management areas contain wood duck boxes as breeding aids.
(Massachusetts Division Fisheries and Game/Jack Swedberg photo)

tour is Delaware Water Gap. Historic Moravian structures may be toured in Bethlehem and Nazareth. Allentown has numerous attractions; among them are the Liberty Bell Shrine, a trout nursery, and tours of Schaefer Brewery (samples are served). The famous Indian athlete, Jim Thorpe, is buried in a town which bears his name. In Hazleton, the Anthracite Museum honors the region's coal industry.

FOR ADDITIONAL INFORMATION:
Public Affairs Officer
U.S. Army Engineer District, Philadelphia
U.S. Custom House
Second and Chestnut Sts.
Philadelphia, PA 19106

Beltzville State Park
Box 252, R.D. 3
Lehighton, PA 18235

SHENANGO RIVER LAKE

Behind Shenango Dam this large reservoir extends north and east through Pennsylvania along the winding path of the Shenango River, spilling over into Ohio through the valley of Pymatuning Creek. Bordered by gently rolling hills, this reservoir forms a U-shaped path approximately 16 miles long. Archeology buffs will find something of special interest awaiting them here. It was determined in a 1951 survey that all prehistoric and historic Indian sites covering five cultural periods, starting in 8000 B.C., are represented in this area; and eight of these sites are on project lands. Of interest from the historic point of view are the well-preserved remains of an Erie Canal stone lock about one mile below the dam. Since several large metropolitan areas lie within easy driving distance of the reservoir, a wide range of recreational facilities have been developed to serve the diversified interests of over a million annual visitors.

HOW TO GET THERE: Follow project signs north from Sharpsville, PA.

FISHING: The fish population includes northern pike, walleye, muskellunge, largemouth bass, bullhead, catfish, suckers, bluegill, sunfish, and crappie. Trout are stocked below Shenango Dam. Special fishing sites are provided for senior citizens and the handicapped. Three launching ramps; lakeside marina, rental boats.

HUNTING: Game animals in the project area include white-tailed deer, squirrel, rabbit, gray fox, duck, geese, pheasant, grouse, and woodcock. A waterfowl propagation area adjacent to the reservoir is managed by the Pennsylvania Game Commission.

CAMPING: Over 225 Corps of Engineers campsites, both primitive and developed; facilities include restrooms, showers, drinking water, tables, fire rings, dump station, and launching ramps. Several private campgrounds in area.

OTHER ACTIVITIES: Swimming, water skiing, canoeing, boating; lakeside marina, boat rentals, three launching ramps. Numerous picnic sites. A section of the old Erie Canal along the Shenango arm of the project has been preserved, and the towpath bordering it is maintained as a hiking trail. The Shenango Valley YMCA, which leases a 43-acre site from the Corps of Engineers, offers swimming, sailing, and golf lessons, as well as junior and senior lifesaving classes, for a nominal fee. In addition, they provide special interest programs during the summer. (For more information, write Shenango Valley YMCA, P.O. Box 925, Sharon, PA 16146.) Near the Upper Mahoney Picnic Area, the Corps has constructed the half-mile-long Seth Myers Nature Trail.

AREA ATTRACTIONS: (Pennsylvania) At the northernmost tip of project lands, across the Shenango River, is King's Mill Bridge; built in 1869, it's the only remaining covered bridge in Mercer County. The Erie National Wildlife Refuge for migratory waterfowl is near Guys Mills. At Meadville you can tour the Allegheny College campus and view a collection of Lincoln materials. The narrow ravine of Oil Creek between Oil City and Titusville was once the busiest valley on the North American continent; over a 10-year period in the late 1800s, over 17 million barrels of oil were extracted from the earth. (Ohio)

Youngstown's Mill Creek Park features a six-mile-long gorge, a formal garden, and the Ford Nature Education Center. In the same city, the Butler Institute of American Art is a showcase for the works of American artists. A beautiful national memorial in Niles commemorates the birthplace of our twenty-fifth President, William McKinley, and houses his personal relics.

FOR ADDITIONAL INFORMATION:
Public Affairs Officer
U.S. Army Engineer District, Pittsburgh
Federal Bldg., 1000 Liberty Ave.
Pittsburgh, PA 15222
Shenango Dam
2442 Kelly Rd.
Sharpsville, PA 16150

YOUGHIOGHENY RIVER LAKE

From its origins within the folds of western Maryland's Allegheny Mountains, the Youghiogheny River flows northward as a wild, rushing stream to the reservoir on the Pennsylvania-Maryland border which bears the stream's name. Here, amid some magnificent mountain scenery, it backs up behind the massive dam for 17 miles, creating 38 miles of twisting shoreline dotted with public recreational facilities. Around the reservoir is a rugged, forest-draped land rich in the early history of our country. The river in both states is wild and beautiful, a stream of cascading falls, remote gorges, and tree-covered banks, offering one of eastern America's most dramatic wilderness experiences.

HOW TO GET THERE: From Confluence, PA, head south on SR 281 for 1.5 mi.

FISHING: Northern pike, walleye, largemouth and smallmouth bass, crappie, and perch provide fishing excitement. A state record pike was snagged here in recent years. The area immediately below the dam is regularly stocked with trout. Over half a dozen launching ramps, some maintained by the Corps of Engineers; lakeside marina, rental boats, supplies.

HUNTING: Deer are abundant here; grouse, squirrel, and raccoon also roam the woods. A word of caution: copperheads are common along the river, and there are rattlesnakes in the higher elevations.

CAMPING: Around 200 Corps of Engineers campsites in several recreation areas; primitive and developed. Facilities include picnic tables, grills, restrooms, drinking water, dump station, and launching ramps. Pennsylvania's largest state park, Ohiopyle, has over 200 modern campsites; it's a few miles north of the dam along the riverbanks. Ice fishing on reservoir in winter.

OTHER ACTIVITIES: Swimming, water skiing, and boating; lakeside marina, rental boats, launching ramps on both sides of the reservoir. The Youghiogheny River is one of the East's finest canoeing streams, though areas near the reservoir are recommended only for those experienced in whitewater. Kayaks and rubber rafts also take to the river. Numerous trails wind through spectacular mountain country; the paths in Ohiopyle State Park are particularly notable. However, hikers should be on the lookout for poisonous snakes throughout the region. Picnic tables around lake. Popular winter sports in surrounding area include snowmobiling, ice skating, and both cross-country and downhill skiing.

AREA ATTRACTIONS: Natural wonders abound in this three-state area (West Virginia is just minutes away). Don't miss Youghiogheny Gorge and Ohiopyle Falls in Ohiopyle State Park, Pennsylvania; Muddy Falls in Maryland's Swallow Falls State Forest; and Cranesville Swamp on the Maryland—West Virginia border north of Terra Alta, WV. The latter is particularly noted for vegetation found nowhere else on the North American continent except the Arctic Circle. (Pennsylvania) Mt. Davis near Savage is the highest point in the Keystone State and is accessible by auto. Near Farmington is Fort Necessity National Battlefield, where George Washington met the first defeat of his military career. Fallingwater, the concrete house which was perhaps Frank Lloyd Wright's most famous design, is northeast of Ohiopyle. (Maryland) Penn Alps near Grantsville features working demonstrations of Colonial arts and crafts. History buffs can retrace our first federal highway, the Old National Road, along U.S. 40 west from Cumberland.

FOR ADDITIONAL INFORMATION:
Public Affairs Officer
U.S. Army Engineer District, Pittsburgh
Federal Bldg., 1000 Liberty Ave.
Pittsburgh, PA 15222
Park Superintendent
Ohiopyle State Park
P.O. Box 105
Ohiopyle, PA 15470

ADDITIONAL CORPS OF ENGINEERS LAKES*

Alvin R. Bush Dam (from Renovo, 5 mi. W on SR 120 to Westport, 8 mi. N on unmarked state road); Conemaugh River Lake§ (follow signs 7 mi. E from Saltsburg); Crooked Creek Lake (from Kittanning, 7 mi. S on SR 66); Curwensville Lake§ (from Clearfield, W on SR 879 to Curwensville, 4 mi. S on SR 453); East Branch Clarion River Lake (9 mi. NE of Johnsonburg); Foster Joseph Sayers Dam§ (from Lock Haven, 11 mi. S on U.S. 220); Francis E. Walter Dam‡, § (from Scranton, S on Pa. Turnpike to Exit 35, E on SR 940); Loyalhanna Lake§ (from Saltsburg, .75 mi. S on SR 981); Mahoning Creek Lake‡, § (follow project signs S from New Bethlehem); Prompton Lake§ (from Scranton, E of U.S. 6 through Honesdale, N on SR 170 to project); Raystown Lake (from Huntingdon, dam road leads E and S for about 6 mi. off U.S. 22); Tionesta Lake (from Tionesta, 1.5 mi. S on SR 36)

SOUTH CAROLINA

CLARK HILL LAKE

Huge, sprawling Clark Hill Lake straddles the states of South Carolina and Georgia, an impoundment of the Savannah River, which forms the boundary line between the two states. The dam lies about 20 miles above Augusta at Clarks Hill, SC, and behind it the reservoir extends 39 miles upstream, spilling over along the way into the scenic valleys of the South Carolina Little River, the Georgia Little River, and the Broad River in Georgia. One of the ten most popular Corps of Engineers lakes in the nation, this year-round playground consistently attracts over 5 million visitors annually. An island-dotted water surface of 72,000 acres is enclosed by nearly 1,200 miles of wooded, twisting shoreline, adjoined in South Carolina by the Sumter National Forest. In addition to these Forest Service lands, numerous Corps of Engineers recreation sites, seven state parks (four in Georgia, three in South Carolina) and many private establishments offer facilities for visitors along the lakeshore. Hickory Knob, South Carolina's first resort state park, offers a lodge complex and cottages. In Georgia, Mistletoe State Park has cottages, while Elijah Clark features both cottages and mobile homes for vacationing guests.

*Unless designated as follows, each project has restrooms, drinking water, and developed campsites. † = no restrooms; ‡ = no drinking water; § = no developed campsites.

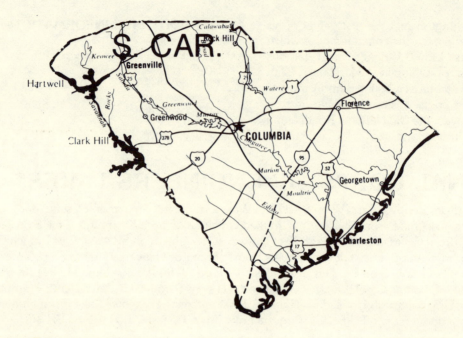

HOW TO GET THERE: From Augusta, GA, take SR 28 northwest for approximately 20 mi.

FISHING: Striped and white bass, striped bass—white bass hybrids, crappie, bream, and catfish are plentiful, but Clark Hill Reservoir is famous for its largemouth bass fishing. There's a walleye population here, but only a few are caught each year. Night fishing with lights is popular summer sport for crappie anglers. Some rainbow and brown trout are stocked below the dam. Over 50 launching ramps on lake; lakeside marinas, boat rentals, supplies. Off the main reservoir body, bass fishing is good during the spring spawning run in the Little River (SC) and in Long Cane and Buffalo creeks branching off this river.

HUNTING: Approximately 33,000 acres of Corps of Engineers project land in both South Carolina and Georgia are open to hunters, as well as parts of Sumter National Forest. Some of South Carolina's largest deer come from this region, and there are special seasons for archery and primitive weapon hunts. Small game includes quail, dove, and duck. There also are some wild turkey; the Bussey Point Wild Turkey

Restoration Area, a cooperative effort, has recently been established on the Georgia side of the reservoir.

CAMPING: Over 300 Corps of Engineers campsites, both primitive and developed. Facilities include restrooms, drinking water, showers, tables, grills, dump stations, and launching ramps. One campground is accessible only to boaters or backpackers. Four state parks in Georgia (Bobby Brown, Elijah Clark, Keg Creek, and Mistletoe) and three in South Carolina (Baker Creek, Hamilton Branch, and Hickory Knob) provide additional campsites at the water's edge; most have water and electrical hookups. There is also camping in Sumter National Forest, but these sites are not on the water. Many private campgrounds in area.

OTHER ACTIVITIES: Lake swimming, water skiing, boating; lakeside marinas, rental boats, over 50 launching ramps. Pools at some state parks. Numerous picnic areas in tree-shaded settings. Lots of opportunity for hiking and nature study. Many rare plants have been identified in the surrounding terrain. This area is rich in history; artifacts abound at several geological sites. Other facilities include bicycling (rental bikes available), horseback riding, and a motorbike trail (Elijah Clark State Park). Several good golf courses in the nearby area.

AREA ATTRACTIONS: (South Carolina) At Greenwood is the only carillon foundry in the United States. The George W. Park Seed Company, just north of Greenwood, offers tours through its greenhouses and experimental gardens. Northwest of McCormick, dating back to 1797, is the John de la Howe School, the oldest manual training foundation in America; visitors welcome. At Bradley is one of three covered bridges left in the state. Price's Grist Mill near Plum Branch is open to the public; it's been providing water-ground cornmeal since 1890. (Georgia) The historical city of Augusta is well worth a visit.

FOR ADDITIONAL INFORMATION:

Public Affairs Officer
U.S. Army Corps of Engineers, Savannah
P.O. Box 889
Savannah, GA 31402

Resource Manager
Clark Hill Project
U.S. Army Corps of Engineers
Clarks Hill, SC 29821

Forest Supervisor
National Forests in South Carolina
1612 Marion St.
Columbia, SC 29201

A fine way to relax and get away from it all—even for dogs! (Arkansas Dept. of Parks and Tourism photo)

HARTWELL LAKE

One of the five most popular Corps of Engineers reservoirs in the nation, Hartwell Lake sprawls across the South Carolina – Georgia border in the heart of some of the East's most beautiful country. From the massive dam near Anderson in the northwestern corner of South Carolina, the vast waters of the reservoir extend along the upper reaches of the Savannah River and into two main branches up the Tugaloo and Seneca rivers. Along the 962-mile shoreline are countless coves and inlets which, even though visitors total over 7 million annually, assure privacy most any time of the year. To the northwest lie the foothills of the Blue Ridge Mountains, a land of towering peaks, cascading waterfalls, and free-flowing streams known to the Cherokee as "The Great Blue Hills of God." Two national forests with a multitude of outdoor recreational opportunities lie just minutes away, while the reservoir itself is surrounded by an excellent highway system and many communities which cater to the needs of tourists.

HOW TO GET THERE: From Anderson, SC, travel south on U.S. 29 for 23 mi.

FISHING: In the main reservoir, there's excellent largemouth bass fishing along the rocky shorelines of numerous islands. During the summer and winter, these fish congregate in the deep holes of the Tugaloo and Seneca, their tributaries, and in the main body of water. A late-winter walleye spawning run up the Tugaloo to Yonah Dam provides good to excellent fishing. Late-summer crappie fishing has proved to be best at night under most highway bridges. Other species include white and hybrid bass, catfish, and bream. Several marinas and over 65 public boat landings; rental boats and supplies at lakeside. The Savannah River for 15 miles below the dam is stocked with rainbow and brown trout. Nearby mountain streams, particularly the Chattooga River, also offer good trout fishing.

HUNTING: Deer and wild turkey are the most popular quarry. Wild hogs may be taken, but are not plentiful. Small game includes grouse, quail, squirrel, rabbit, and raccoons. Several wildlife management areas in the Chattahoochee National Forest in Georgia and Sumter National Forest in South Carolina. Also some good small game hunting around the reservoir.

CAMPING: Over 500 Corps of Engineers campsites, both primitive and developed. Facilities include restrooms, showers, drinking water, tables, grills, firewood, and launching ramps. In addition, there are campsites with water and electrical hookups at three lakeshore state parks—Tugaloo and Hart in Georgia and Sadlers Creek in South Carolina. Chattahoochee National Forest in Georgia and Sumter National Forest in South Carolina offer campgrounds in lovely, secluded spots, but they are away from the main reservoir. Hartwell Lake is one of the few places in the country where campers may sample the unique Camp-A-Float cruisers. Self-contained recreational vehicles are secured to rental pontoon boats to create a totally private, floating campsite. (For more information, write Camp-A-Float, P.O. Box 1625, Rockford, IL 61110.)

OTHER ACTIVITIES: Lake swimming, skin and scuba diving, water skiing, boating; lakeside marinas, rental boats and supplies, over 65 launching ramps. Canoeing on the wild and lovely Chattooga River, where the movie *Deliverance* was filmed (some stretches definitely for experts only). Outdoor enthusiasts will find a vast variety of activities, including horseback riding, picnicking, rockhounding, nature study, and hiking. An exhibit in the office of Lake Hartwell's resource manager points out area attractions. A portion of the Appalachian Trail winds through Chattahoochee National Forest; another opportunity for backpacking exists in Ellicott's Rock Wilderness Area in Sumter National Forest. Scenic areas abound in the national forests, and many may be reached by auto. North of Hartwell Lake is SR 11, the "Cherokee Foothills Scenic Parkway," which offers panoramic mountain scenery.

AREA ATTRACTIONS: (Georgia) Atop Chenocetah Mountain near Cornelia is a tower which permits a view of four states on clear days. Tallulah Gorge Park, near Tallulah Falls, features a deep gorge with waterfalls and lush vegetation. A drive northeast of Toccoa on SR 184 takes motorists across the Tugaloo River via a picturesque covered bridge. (South Carolina) Stump House Mountain tunnels, near Walhalla, are used to cure blue cheese. Also at Walhalla is a national fish hatchery; visitors are welcome. Attractions at Clemson include Clemson University's National Horticultural Ornamental Gardens and the Keowee-Toxaway Visitor's Center, which depicts the story of electricity. History buffs will enjoy the Pendleton Historic District of old homes. Bob Jones University at Greenville is noted for its Museum of Sacred Art.

FOR ADDITIONAL INFORMATION:
Public Affairs Officer
U.S. Army Engineer District, Savannah
P.O. Box 889
Savannah, GA 31402

Resource Manager
Hartwell Lake Project
P.O. Box 278
Hartwell, GA 30643

Forest Supervisor
National Forests in Georgia
Box 1437
Gainesville, GA 30501

Forest Supervisor
National Forests in South Carolina
1612 Marion St.
Columbia, SC 29201

SOUTH DAKOTA

GAVINS POINT DAM (LEWIS AND CLARK LAKE)

The natural beauty of clear blue waters, the dramatic effect of the chalk cliffs along its northwestern shoreline and the rolling, wooded hills of the prairie which surrounds it create the setting which makes Lewis and Clark Lake one of the most popular Corps of Engineers reservoirs. The smallest and most scenic of South Dakota's Great Lakes (four Corps of Engineers impoundments along the Missouri River), Lewis and Clark Lake, formed by Gavins Point Dam, straddles the southeastern border of the Sunshine State and spills over into Nebraska. The history which passed this way along the banks of the Missouri River included the powerful Sioux Indian nation, the Lewis and Clark expedition, the notorious James brothers, who once used this area as a hideout, and Custer's Seventh Cavalry, which paused at Yankton on its way to the Little Big Horn. Today's visitors come to seek unsurpassed water-based recreation on a 40-mile-long reservoir with facilities to fit every desire.

HOW TO GET THERE: From Yankton, SD, head west on SR 52.

FISHING: Anglers fill their stringers with walleye, sauger, catfish, pike, bass, crappie, paddlefish, and a variety of pan and rough fish. In a single season the tailwaters below the dam have put out more than half a million pounds of sport fish. State records abound in both South Dakota and Nebraska. Winter fishing is

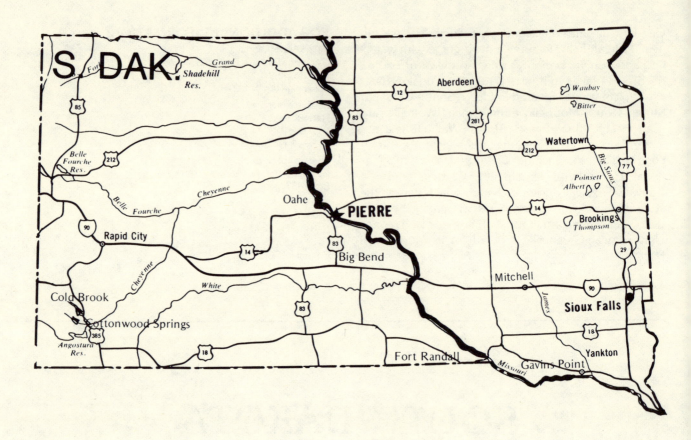

done in the tailwaters of the dam, which usually don't freeze, while on the reservoir itself villages of ice houses are common. Over 15 launching ramps; lakeside marina, rental boats, supplies.

HUNTING: Deer is the only big game here, with both whitetails and mulies present. Numerous migrating ducks and geese pause here. Pheasants, wild turkey, sharptail grouse, and rabbit also live here. Designated hunting areas in both South Dakota and Nebraska.

CAMPING: Over 250 Corps of Engineers campsites, both developed and primitive. Facilities include restrooms, showers, drinking water, tables, grills, some electrical hookups, dump stations and launching ramps. Both South Dakota and Nebraska maintain state recreation areas on the shores of the lake, with a total of over 350 sites for camping.

OTHER ACTIVITIES: Swimming, water skiing, skin and scuba diving, pleasure boating of all types; lakeside marinas, rental boats, over 15 launching ramps. Youngsters of all ages enjoy building rafts from drift-wood here. An ideal spot for sailboating; canoeing on the Missouri River from Fort Randall Dam to Lewis and Clark Lake and on the Niobrara River. Take off for a hike or plan a leisurely picnic lunch. There's horseback riding at Nebraska's Niobrara State Park near the town of the same name. The area is rich in fossils, and some good finds have been made around the lakeshore. There are scenic drives along the Missouri River. Wildlife to be observed includes fox, coyote, mink, weasel, muskrat, striped and spotted skunk, beaver, wild turkey and bobcat; it's a good place for birdwatching, too, as the northern bald and golden eagle winter near the tailwaters below Gavins Point Dam. Below Fort Randall Dam upstream is a bald eagle refuge which is a Registered National Landmark; it's cosponsored by the Corps of Engineers and the National Wildlife Federation. Many interesting rock formations in area. Ice skating in winter.

AREA ATTRACTIONS: Visitors are welcome at a national fish hatchery below Gavins Point Dam. Also near the dam are a Lewis and Clark campsite marker and historical display at project headquarters. At the upper end of the lake, near a town with the lyrical name of Running Water, is one of the Missouri River's few operating ferries; it's been plying these waters since 1875, but does so on a seasonal basis. The Santee Indian Reservation borders part of the reservoir's Nebraska shoreline. On the University of South Dakota campus at Vermillion is a museum which features exhibits of archeological and historical interest, as well as a collection of antique musical instruments.

FOR ADDITIONAL INFORMATION:

Public Affairs Officer
U.S. Army Engineer District, Omaha
7410 U.S. Post Office and Court House
215 N. 17th St.
Omaha, NE 68102

Reservoir Manager
Lewis and Clark Lake
Yankton, SD 57078

Great Lakes of South Dakota Association
Box 786
Pierre, SD 57501

LAKE OAHE

The gargantuan prairie giant that is Lake Oahe sprawls north and south through the historic Missouri River Valley between Pierre, SD, and Bismarck, ND, creating huge embayments along the way that are lakes in themselves. Largest of South Dakota's Great Lakes (four vast Corps of Engineers impoundments on the Missouri River), Oahe offers 313,000 surface acres which touch 2,250 miles of shoreline. South Dakota claims more boats per capita than any other state, and Lake Oahe is one big reason why. Nearly all recreation facilities here are Corps-managed.

HOW TO GET THERE: From Pierre, SD, head 8 mi. north on SR 514.

FISHING: The walleyes and northern pike have attracted anglers from all over the country. Other species available are white bass, channel catfish, white and black crappie, yellow perch, bluegill, sauger, and paddlefish. In the winter, ice fishing houses dot the reservoir's surface. Nearly 20 launching ramps; lakeside marina, rental boats, supplies.

HUNTING: All the Great Lakes are known for their waterfowl hunting, primarily mallards and Canada geese. Draws along the lake offer grouse, Hungarian partridge, pheasant, deer, and antelope.

CAMPING: Nearly 500 Corps of Engineers campsites, both primitive and developed. Facilities include restrooms, showers, picnic tables, grills, drinking water, some electrical hookups, dump stations, and launching ramps. In addition, around 100 campsites are provided by various local governments around the reservoir.

OTHER ACTIVITIES: Swimming, water skiing, skin and scuba diving, boating of all types; lakeside marina, rental boats. Good winds for sailing. The Missouri River from Riverdale to Bismarck is the most popular canoe route in North Dakota; from Bismarck to Oahe Dam is recommended for experts only. Lots of room for hiking or picnicking. Rockhounding is great in both South Dakota and North Dakota. Many interesting rock formations for nature study; among the wildlife visitors may see are bobcat, snakes (including the prairie rattler), deer, raccoon, muskrat, beaver, hawks, and waterfowl. Ice skating in winter.

AREA ATTRACTIONS: Two Indian reservations border Lake Oahe's western shore—Cheyenne River Reservation in South Dakota and Standing Rock Reservation in North Dakota. Famed Indian Chief Sitting Bull was killed near Little Eagle, SD; his body was first buried near Fort Yates, ND, then later moved to its present hilltop site above the Missouri River Valley near Mobridge, SD. Murals depicting the history of area Indians, done by a South Dakota Sioux artist, may be viewed at Mobridge Municipal Auditorium. Fort Lincoln State Park near Mandan, ND, once the headquarters for Custer's ill-fated Seventh Calvary, features a restored Slant Indian Village. Visitors are welcome at the Northern Great Plains Field Station near Mandan; a protective forestry station which specializes in windbreaks and shelterbelts, it's one of the largest in the United States. Long Lake National Wildlife Refuge is near Moffit, ND; sandhill cranes use this area during migration. The Flaming Fountain on South Dakota's capitol grounds at Pierre is a spectacular nighttime sight; flames reflecting the colors of the rainbow leap from natural artesian waters.

FOR ADDITIONAL INFORMATION:

Public Affairs Officer
U.S. Army Engineer District, Omaha
7410 U.S. Post Office and Court House
215 N. 17th St.
Omaha, NE 68102

Area Engineer
Oahe Reservoir
P.O. Box 196
Pierre, SD 57501

Great Lakes of South Dakota Association
Box 786
Pierre, SD 57501

ADDITIONAL CORPS OF ENGINEERS LAKES*

Big Bend Dam, or Lake Sharpe (from Sioux Falls, W on I-90 to Chamberlain, N on SR 47); Cold Brook Lake‡ (about 1 mi. N of Hot Springs); Fort Randall Dam, or Lake Francis Case (near Pickstown, on U.S. 18)

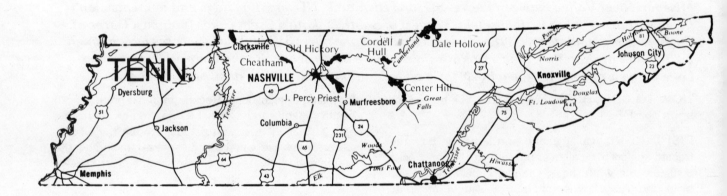

TENNESSEE

CENTER HILL LAKE

Situated in the green and rolling foothills of the Cumberland Mountains of central Tennessee, Center Hill Lake creates 415 miles of wooded shoreline in the valley of the Caney Fork River. The Volunteer State is noted for its lovely waterfalls, and three of them—Fanchers Falls, Great Falls, and Burgess Falls—are located near this reservoir. A multi-fingered lake, Center Hill finds its way into many picturesque creeks along its 64-mile path. In addition to several Corps of Engineers recreation areas, Rock Island State Park and Edgar Evins State Rustic Park border the water. Future development at Edgar Evins, one of Tennessee's newest parks, will include a number of cabins, a swimming pool, and tennis courts.

HOW TO GET THERE: From Cookeville, head west on I-40 for 14 mi.; turn south for 1 mi. on SR 56, then west on SR 141 for about 8 mi. to the dam.

FISHING: Smallmouth bass are probably the most sought-after fish in Center Hill Lake, noted also for its largemouth bass, crappie, and walleye population.

---

*Unless designated as follows, each project has restrooms, drinking water, and developed campsites. † = no restrooms; ‡ = no drinking water; § = no developed campsites.

The tailwaters are excellent for rainbow and brown trout fishing; float fishing is popular on the Caney Fork River below the dam. Nearly a dozen launching ramps; lakeside marina, rental boats, supplies.

HUNTING: Dove, squirrel, quail, grouse, rabbit, and deer inhabit the vast woodlands around the reservoir.

CAMPING: Nearly 250 developed Corps of Engineers campsites; facilities include restrooms, drinking water, showers, tables, grills, and launching ramps. Additional campsites at two lakeshore state parks. Some private docks around the lake offer trailer spaces with hookups.

OTHER ACTIVITIES: Several swimming beaches; a pool is planned at the new Edgar Evins State Rustic Park. Water skiing and boating; lakeside marina, rental boats, including houseboats. Nearly a dozen launching ramps. Canoeing on Caney Fork River below the dam or on the Obed River near Crossville. Nature trails and picnicking. Tennis courts are being constructed at Evins State Park.

AREA ATTRACTIONS: Fall Creek Falls State Park west of Pikeville features some of the most spectacular scenery in the state, and you can view it from foot or horseback. The falls for which the park is named is the highest east of the Rockies. At Old Stone Fort Archeological Area near Manchester is yet another waterfall and an ancient walled structure nearly 2,000 years old; its origin remains a mystery to this day. Those who like to do their exploring underground will enjoy Cumberland Caverns near McMinnville.

FOR ADDITIONAL INFORMATION:

Public Affairs Officer
U.S. Army Engineer Officer, Nashville
P.O. Box 1070
Nashville, TN 37202

Reservoir Manager
Center Hill Dam
Lancaster, TN 38569

DALE HOLLOW LAKE

Winding through the mountains and valleys of the Cumberland foothills of north central Tennessee, spilling over into Kentucky, is beautiful Dale Hollow Lake. The numerous small islands which dot the 28,000-acre reservoir, along with many coves and inlets which penetrate its 620-mile shoreline, provide privacy aplenty for those who want to get away from it all. Because the forested, mountainous region which surrounds Dale Hollow permits only scattered farming, and because the population density has remained low, the water here is exceptionally clear. This reservoir is one of the oldest in the area, and its shoreline is dotted with well-established vacation facilities. Standing Stone State Park, near the Tennessee shoreline of the lake, offers a lodge, cabins, timberlodges, group lodges, and a restaurant; Dale Hollow Lake State Park is under construction in Kentucky.

HOW TO GET THERE: From Livingston, TN, follow SR 52 north for 20 mi. to Celina; then follow signs to dam.

FISHING: Known far and wide as one of the best fishing lakes in the country. The 11 lb., 15 oz. world record smallmouth was taken from Dale Hollow in 1955, and many believe if there is a larger one around it will come from these same waters. Anglers will also find abundant largemouth and white bass, bluegill and crappie, as well as rainbow trout and an occasional muskie. Nearly a dozen launching ramps; lakeside marinas, rental boats, supplies.

HUNTING: Along the shoreline of Dale Hollow, hunters stalk deer, squirrel, rabbit, and raccoon. The lake itself serves as a resting place for migratory waterfowl.

CAMPING: Nearly 250 developed Corps of Engineers campsites; facilities include restrooms, drinking water, showers, tables, grills, and launching ramps. Additional campsites available in Standing Stone State Park (TN) and Dale Hollow Lake State Park (KY), now under construction but soon to be completed. Many private campgrounds in area.

OTHER ACTIVITIES: Swimming in lake and at the Standing Stone State Park pool; skin and scuba diving, water skiing, boating; nearly a dozen launching ramps, lakeside marinas, rental boats (including houseboats). Scenic picnic areas. Many hiking trails in vicinity. A good spot for nature study. Horseback riding allowed in undeveloped areas. Canoeing, backpacking, and scenic drives in nearby Daniel Boone National Forest.

AREA ATTRACTIONS: (Kentucky) The Old Mulky Meetinghouse State Shrine near Tompkinsville is of great historic significance in this area. North of Burkesville is the site of the first American oil well. (Tennessee) A national fish hatchery below Dale Hollow Dam is open to the public. One particularly scenic drive is SR 52 east from Livingston through Jamestown to its junction with U.S. 27; visitors pass through rugged "upper Cumberland country." Pickett State Park northeast of Jamestown offers a wilderness of notable geological formations and a variety of vegetation. A scenic trail leads to the nearby home and grist mill of World War I hero, Sgt. Alvin York.

FOR ADDITIONAL INFORMATION:

Public Affairs Officer
U.S. Army Engineer District, Nashville
P.O. Box 1070
Nashville, TN 37202

Resource Manager
Dale Hollow Dam
P.O. Box 276
Celina, TN 38551

Standing Stone State Park
State Hwy. 52
Livingston, TN 38570

Dale Hollow Lake State Park
c/o Kentucky Dept. of Parks
Capital Plaza Tower
Frankfort, KY 40601

J. PERCY PRIEST RESERVOIR

Just east of Nashville, TN, and south of I-40 lies J. Percy Priest Dam. The sprawling reservoir it impounds stretches up the Stones River, its waters dotted with islands, its hilly shoreline dotted with trees. Three interstate highways converge at Nashville, and the accessibility they provide, as well as the lake's nearness to metropolitan Nashville, contribute to the tremendous popularity enjoyed by Priest Reservoir. Though new, it ranks among the Corps of Engineers' 10 most popular recreation areas, luring nearly 5 million visitors yearly to its scenic shores. A panoramic view of the lake is available from the award-winning Visitor Center.

HOW TO GET THERE: From Nashville, head east 6 mi. on I-40 to the Hermitage exit; then turn right on Old Hickory Blvd. for 1 mi.

FISHING: A rockfish stocking program has proved tremendously successful. Bass, both largemouth and smallmouth, bream, and crappie are other popular species. Good sauger fishing in the tailwaters below the dam. Over 25 launching ramps; lakeside marina, rental boats, supplies.

HUNTING: A state wildlife management area adjoining the lake provides hunting for waterfowl, quail, dove, rabbits, squirrels, and a few deer.

CAMPING: Nearly 500 developed Corps of Engineers campsites. Facilities include restrooms, drinking water, showers, tables, grills, and launching ramps. Cedars of Lebanon State Park, about five miles east of the reservoir, has several modern campsites, as well as a group camping area.

OTHER ACTIVITIES: Lake swimming, Olympic-size pool and wading pool at Cedars of Lebanon State Park; water skiing, boating. Over 25 launching ramps; lakeside marinas, rental boats. Also picnicking. Cedars of Lebanon State Park features square dancing every Saturday night, nature trails, and horseback riding.

AREA ATTRACTIONS: Many lovely antebellum homes in area. The largest remaining red cedar forest in the United States may be seen at Cedars of

Streams leading into some of the lakes offer a chance of quiet escape. (Dave Murrian/TWRA photo)

Lebanon State Park. Stones River National Battlefield near Murfreesboro offers "living history" demonstrations during the summer; there's also a national environmental study area and a national cemetery there. The city of Nashville offers many attractions; among them are the Grand Ole Opry, Opryland U.S.A. (a huge family amusement park), and the Country Music Hall of Fame.

FOR ADDITIONAL INFORMATION:

Public Affairs Officer
U.S. Army Engineer District, Nashville
P.O. Box 1070
Nashville, TN 37202

Reservoir Manager
J. Percy Priest Resident Officer
P.O. Box 2347
Nashville, TN 37214

OLD HICKORY RESERVOIR

This long, narrow, twisting lake snakes around numerous bends as its waters back up the Cumberland River behind Old Hickory Lock and Dam in north central Tennessee. In the spring its upper reaches come alive with beauty when the redwood and dogwood burst into bloom, while all along the 400-mile shoreline cedar and mixed hardwoods are abundant. With metropolitan Nashville less than 25 miles away and with I-65 passing near the dam, Old Hickory Lake enjoys enormous popularity, attracting over 5 million visitors annually and ranking as one of the Corps of Engineers' 10 most popular reservoirs.

HOW TO GET THERE: From Nashville, take I-65 north to the Madison exit (SR 45); then turn east about 5 mi. to the dam.

FISHING: Prime walleye waters; the world's record, a 25-pounder, came from Old Hickory waters. Anglers will also find largemouth bass, crappie, bream, catfish, white bass, and an occasional rockfish. Sauger fishing is particularly good in the tailwaters. Nearly 25 launching ramps; marinas, rental boats, supplies at lakeside.

HUNTING: The state has established wildlife refuges and management areas adjacent to the lake. A permanent resident population of over 500 Canadian geese live in the area. Also duck, doves, rabbits, quail, and squirrels. A few deer in the upper reaches of the reservoir.

CAMPING: Nearly 75 developed Corps of Engineers campsites; facilities include restrooms, drinking water, showers, tables, grills, launching ramps. Bledsoe Creek State Park, as well as several private agencies along the lakeshore, provide many additional campsites with full hookups.

OTHER ACTIVITIES: Lake swimming, pool at Bledsoe Creek State Park; water skiing, boating. Nearly 25 launching ramps; lakeside marinas, rental boats, and canoes. Many picnic areas. Nature trails and photography blinds.

AREA ATTRACTIONS: Numerous antebellum homes in area. The Hermitage, Andrew Jackson's magnificent home not far from the lakeshore, may be toured. Nashville offers a multitude of attractions. You may attend the Grand Ole Opry or visit Opryland U.S.A., a huge family amusement park. There's a Country Music Hall of Fame and Museum. The Parthenon is a full-size replica of its famous namesake in Athens, Greece. There are scenic drives, as well as bridle and hiking trails, through Percy and Edwin Warner parks. The *Captain Ann,* a dual stern paddlewheel boat, will take you aboard for a cruise down the Cumberland River. Springfield offers tobacco auctions from December to March.

FOR ADDITIONAL INFORMATION:

Public Affairs Officer
U.S. Army Engineer District, Nashville
P.O. Box 1070
Nashville, TN 37202

Reservoir Manager
Old Hickory Resident Office
P.O. Box 511
Old Hickory, TN 37138

Floating and fishing on the Buffalo River. (Dave Murrian/TWRA photo)

ADDITIONAL CORPS OF ENGINEERS LAKES*

Cheatham Lake (from Nashville, 32 mi. W on SR 12); Cordell Hull Reservoir (from Nashville, E on I-40 to
Carthage exit, N on SR 80 to Carthage)

*These projects have restrooms, drinking water, and developed campsites.

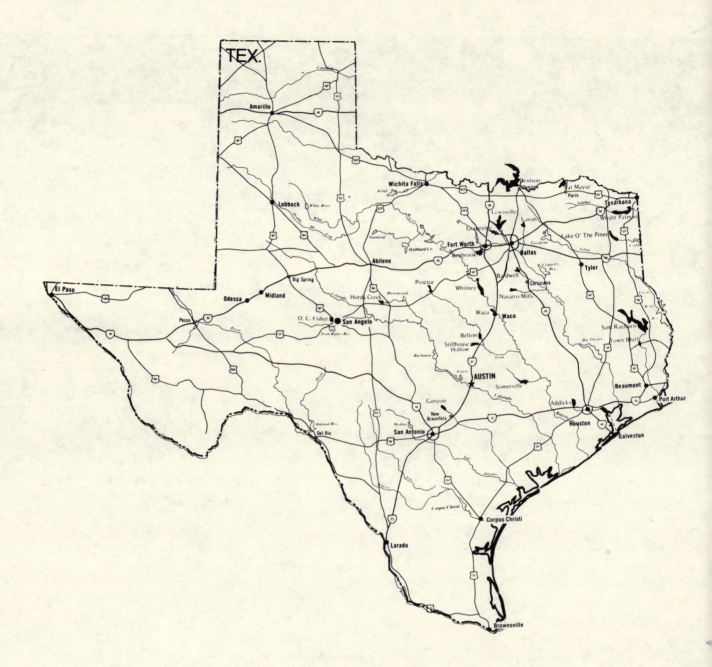

TEXAS

DENISON DAM (LAKE TEXOMA)

Wide expanses of blue water, a lush green shoreline, and an unlimited variety of activities have combined to make Lake Texoma, formed by Denison Dam, the second most popular Corps of

Engineers reservoir in the country. Sprawling along the Oklahoma-Texas border in fabled Red River country, it annually attracts more than 13 million visitors. Two state parks (one with a luxurious lodge), two national wildlife refuges, plush private resorts, private cottages, and many campgrounds are scattered along its 680 miles of shoreline. Enriching the pioneer history of this land were men like President Zachary Taylor, General Robert E. Lee, Davy Crockett, and Jim Bowie, all of whom visited old Fort Washita, which still stands today near the water's edge. Years later, in Denison, Texas, a child was born who was destined to lead his country as both a soldier and a president; Dwight D. Eisenhower State Park on Lake Texoma honors his memory, and his birthplace is open to the public. Despite Texoma's popularity, there are some getaway spots here; hiking trails, isolated coves and inlets, several islands, and two wildlife refuges offer quiet havens away from developed areas.

A beach at Lake Texoma on the Texas-Oklahoma border.

HOW TO GET THERE: From Denison, Texas, head 5 mi. NW on SR 75A.

FISHING: Some of the best fishing in the nation is found here. White bass, black bass, and crappie are the primary species, with channel, blue, and flathead catfish, sunfish, and bluegill also providing action. The striped bass was introduced here in 1969, and some in excess of 25 pounds are being caught in the lake and its tailwaters. Walleye were introduced more recently. Rainbow trout fishing in nearby Blue River is an added bonus. More than 80 public launching ramps; lakeside marinas, rental boats, supplies. Fishing comfort is provided by several enclosed docks which are air-conditioned in summer and heated in winter.

HUNTING: Texoma is a prime area for waterfowl. Deer and quail hunting are also very popular. Other principal game species are wild turkey, dove, cottontail rabbit, and squirrel. Approximately 50,000 acres of project land are open for hunting, and there are many other public hunting areas nearby.

CAMPING: The Corps of Engineers operates well over 1,000 campsites, both primitive and developed. Facilities include restrooms, showers, drinking water, tables, grills, dump stations, and launching ramps. Over 650 more sites are available in Eisenhower State Park (TX) and Lake Texoma State Park (OK). Innumerable private campgrounds skirt the lake. Every facility imaginable is available here. Some island campsites are accessible by boat.

OTHER ACTIVITIES: Lake and pool swimming, water skiing, boating of all types; lakeside marinas rent boats, and there are more than 80 public launching ramps. Picnic areas all around lake. Beachcombing may result in finds of arrowheads, marine fossils of prehistoric times, or pieces of petrified wood; visi-tors may learn about area fossils by checking the paleontology exhibits at the Denison Dam power-house. Bird-watchers please note: more than 500 species of birds have been logged at Texoma, including the rare whooping crane. Tishomingo and Hagerman national wildlife refuges are good spots for nature study. The 14-mile-long Cross Timbers Hiking Trail leads hikers across steep rocky ledges, into dense woodlands, and through many changes of elevation. Texoma State Park offers horseback riding, archery, bicycling, skeet shooting, tennis and other game courts; golfers will find a nine-hole course, putting green, and driving range. Many annual festivals and events are held nearby.

AREA ATTRACTIONS: (Oklahoma) In Tishomingo are the last capitol of the old Chickasaw nation and an adjacent museum; tribal business is still conducted in this city. Nearby are the Tishomingo National Fish Hatchery and Tishomingo National Wildlife Refuge, primarily for migrating waterfowl; both welcome visitors. Our country's smallest national park, Platt, is near Sulphur; it offers nationally famous mineral springs, a small buffalo herd, and the Travertine Nature Center. The Arbuckle National Recreation Area in the Arbuckle Mountains southwest of Sulphur features rocky bluffs, springs, and clear-running streams, as well as diverse and unusual rock formations. Visitors are welcome at the Samuel Roberts Noble Foundation in Ardmore, which engages in medical and agricultural research. (Texas) Sam Rayburn Library in Bonham is filled with mementos of the noted congressman's life.

FOR ADDITIONAL INFORMATION:
Public Affairs Officer
U.S. Army Engineer District, Tulsa
P.O. Box 61
Tulsa, OK 74102
Lake Texoma Association
Box 1128
Denison, TX 75020

GRAPEVINE LAKE

Surrounded by populous cities and an excellent highway system in northeastern Texas, Grapevine Lake is one of the most popular reservoirs in the Lone Star State. Its waters, as they back up among the gently rolling hills which border Denton Creek, create a 60-mile shoreline. To accommodate the more than 4 million visitors who annually come here, the Corps of Engineers has established several recreation areas at the water's edge. Fort Worth and Dallas, each less than 25 miles away, offer many cultural attractions which add yet another dimension to a visit to this reservoir.

HOW TO GET THERE: From Grapevine, head 2 mi. NE on SR 121.

FISHING: The largemouth bass is king here, with white bass running a close second. Catfish and crappie are also found here. Over a dozen launching ramps; lakeside marina, rental boats, supplies.

HUNTING: Deer is the primary target for big game hunters here, while squirrel and quail are the most popular small game. Hunting permitted on designated reservoir lands.

CAMPING: Over 200 Corps of Engineers campsites, both primitive and developed. Facilities include restrooms, drinking water, showers, tables, grills, and launching ramps. Several private campgrounds in area.

OTHER ACTIVITIES: Swimming, water skiing, boating; lakeside marina, rental boats, over a dozen launching ramps. Many picnic areas. Denton Creek below the dam is a popular float stream when water levels permit. Golf, tennis, and horseback riding available nearby.

AREA ATTRACTIONS: The people of Dallas have erected a John F. Kennedy Memorial near the site of the late president's assassination. Over 4,000 fish and many species of amphibians inhabit the aquarium in Dallas' State Fair Park. Two drive-through wild animal parks—World of Animals and Lion Country Safari—are near Dallas. The Amon Carter Museum of Western Art at Fort Worth houses an outstanding collection of paintings and sculptures by some of the West's most famous artists. Forest Park in the same city offers rides aboard one of the longest miniature railroads in the United States. North of Fort Worth, nature trails lead visitors through the Greer Island Nature Center and Refuge. Six Flags Over Texas near Arlington is a vast family amusement park which is one of the state's most popular tourist attractions. Cauble Ranch near Denton features a working area dedicated to Cutter Bill, the world champion cutting horse of 1962. The North Texas State University Museum provides an interesting glimpse into Texas history.

FOR ADDITIONAL INFORMATION:
Public Affairs Officer
U.S. Army Engineer District, Fort Worth
P.O. Box 17300
Fort Worth, TX 76102

Grapevine Lake Manager
Box 326
Grapevine, TX 76051

LAKE O' THE PINES

Natural scenic beauty and varied recreational activities combine to make Lake o' the Pines one of the most beautiful man-made lakes in all of Texas. It is so broad one can hardly see one shoreline from another. Recreation areas are well spaced around the lake, with good roads linking them for visitor convenience. More than a dozen commercial concession facilities are located on the lake, and a like number are operated by the Corps of Engineers. The lake is a mix of cleared areas for safe boating and water skiing and vast acreage where the timber was left standing to promote excellent fishing waters. Waterfowl, too, use these wooded areas extensively during some seasons of the year. A normal water surface area of about 18,000 acres provides some excellent habitat for wildlife along the lakeshore.

HOW TO GET THERE: From Jefferson, head west for 9 mi. on SR 49, then continue west on FM 729 and FM 726.

FISHING: Largemouth and smallmouth bass, bream, crappie, and bluegill are the main species found here. Some eight-pound largemouth have been taken from the wooded portions of the lake. Best crappie fishing is in the months of April and May and again in October. Bass fishing is best from February through May and again during the late fall months. More than 25 launching ramps; lakeside marinas, rental boats, supplies.

HUNTING: Waterfowl hunting, especially for ducks, is excellent during the fall months up until December in this area. Many ducks winter on this lake; Canada geese also use the lake. On land hunters find excellent squirrel, rabbit, and partridge hunting.

CAMPING: Almost 500 Corps of Engineers campsites, both primitive and developed. Facilities include restrooms, drinking water, showers, tables, grills, electrical hookups, and launching ramps. Several commercial concessionaires along the water's edge also offer camping with varying facilities. Private campgrounds in area.

OTHER ACTIVITIES: Swimming, water skiing, boating; more than 25 launching ramps, lakeside marinas, rental boats. Canoeing is becoming increasingly popular on Lake o' the Pines.

AREA ATTRACTIONS: Caddo Lake is one of the most unusual of the area attractions; many claim it was formed by the same 1811 earthquake that created Reelfoot Lake in Tennessee. It has a wilderness, swamplike atmosphere with dense stands of cypress trees throughout the lake. A state park is located there, and there are many fine restaurants and overnight accommodations. Marshall has many unusual attractions, including a unique boat-building operation where Louisiana-type electric pirogues named Feather Craft are made and one of the nation's largest pottery factories. At Jefferson a restoration of the beautiful and historic downtown area is underway, with many structures bearing Texas state historical medallions. The many oilfields near Longview may be toured.

FOR ADDITIONAL INFORMATION:

Public Affairs Officer
U.S. Army Engineer District, New Orleans
P.O. Box 60267
New Orleans, LA 70160

Reservoir Manager
Lake o' the Pines
P.O. Drawer "W"
Jefferson, TX 75657

Sandy beaches and diving floats provide fun and relaxation for everyone.

LEWISVILLE LAKE

Located on the Elm Fork of the Trinity River some 22 miles northwest of Dallas, TX, Lewisville Lake, officially named the Garza—Little Elm Reservoir, backs up an impoundment of 23,280 surface acres with a shoreline of 183 miles. Its many secluded inlets and coves harbor largemouth bass, crappie, and bluegill, caught mostly during the spring months. But fishing is excellent on a year-around basis, and sizeable catches are made during all seasons. Being close to the large population centers of Dallas, Denton, and Arlington, it naturally gets much use, but the lake is so vast that seldom is one unable to find seclusion and privacy.

HOW TO GET THERE: From Dallas, take I-35E north to Lewisville, which is located near the southwest end of the lake.

FISHING: Good catches of largemouth bass, crappie, and bluegill, as well as channel catfish, are made here year after year. Spring and fall are the best seasons for the big ones. Summertime temperatures become so hot the fishing falls off considerably. Some areas where the timber was left standing provide excellent bass habitat. East Hill Park features a fishing barge. More than 30 launching ramps; lakeside marina, rental boats, supplies.

HUNTING: Deer, dove, quail, and some types of waterfowl are the main target of hunting activity in this area.

CAMPING: More than 150 Corps of Engineers campsites, both primitive and developed. Facilities include restrooms, drinking water, showers, tables, grills, electrical hookups, and launching ramps. Several local communities also operate recreation areas with campgrounds at the water's edge. Lewisville Park has a dump station. A number of private campgrounds, too, are located within an hour's drive of the lake, providing ample space for all comers.

OTHER ACTIVITIES: Swimming, water skiing, boating; more than 30 launching ramps, lakeside marina, rental boats. Lewisville Park offers a golf course, playground, and bicycle trail. All recreation areas provide picnic facilities. Some canoeing is done on the Elm Fork of the Trinity River.

AREA ATTRACTIONS: Texans will tell you Dallas has everything; the tallest buildings in the South, the state fair of all state fairs each October and, on the First National Bank Building's fifth floor, a unique money tree made from more than 7,800 rare and common coins and polished woods. It also has the glamorous Market Center, Six Flags Over Texas (a Disneyland-type of operation) and Nieman-Marcus, perhaps the most famous department store in the world. Fort Worth, just west of Dallas, also has numerous attractions worth exploring. At Denton the Texas Agriculture Experiment Station; the Cauble Ranch, a working ranch which has produced a world champion cutting horse; and the North Texas State University Museum with its collections of Texana may be toured.

FOR ADDITIONAL INFORMATION:

Public Affairs Officer
U.S. Army Engineer District, Fort Worth
P.O. Box 17300
Fort Worth, TX 76102

Project Engineer
Garza—Little Elm Reservoir
P.O. Box 248
Lewisville, TX 75067

O. C. FISHER DAM (SAN ANGELO LAKE)

In the semiarid ranch country of west central Texas is an oasis called San Angelo Lake, impounded from the waters of the North Concho River by O. C. Fisher Dam. At normal pool, the lake extends to a length of more than eight miles, reaching into the mouths of several creeks along the way to create secluded coves and inlets for fishermen. It boasts a 27-mile shoreline along which some bragging-size largemouth bass are caught the year around. This is excellent game country, too. In the near-desert region which surrounds the reservoir, brilliant desert blooms add touches of brightness to the landscape. Part of the area adjoining the lake is used for agricultural research and wildlife management. Recreational facilities are operated by the Corps of Engineers and the city of San Angelo.

HOW TO GET THERE: The lake is adjacent to the western city limits of San Angelo, off FM 853 and FM 2288.

FISHING: Largemouth bass, bluegill, channel catfish, perch, and crappie are the main species of fish found here. Best seasons are in the spring, from March until May, and again during the autumn months of September to November. About 20 launching ramps.

HUNTING: Wild turkey, white-tailed deer, javelina, upland game birds, and migratory waterfowl provide some excellent hunting.

CAMPING: Over 100 Corps of Engineers campsites, both primitive and developed; facilities include restrooms, drinking water, tables, grills, dump stations, electrical hookups, and launching ramps. Private campgrounds in the area include a KOA with full hookups in San Angelo.

OTHER ACTIVITIES: Swimming, boating; about five launching ramps. Picnicking, hiking, and trailbiking are other popular activities here. The Corps of Engineers maintains a special area for off-road vehicles. There's an 18-hole golf course operated by the city of San Angelo in Riverside Park adjoining the reservoir.

AREA ATTRACTIONS: Old Fort Concho in San Angelo features several restored buildings of a typical frontier Indian fort, as well as a museum. Tourists enjoy a ghost town located near the Old Fort Chadbourne Ruins close to Bronte. The vast Caverns of Sonora, near the city of Sonora, offer a profusion of formations in a full range of colors. These caves were described by a past president of the National Speleological Society as "the most indescribably beautiful in the world. . . . [Their] beauty can't possibly be exaggerated—even by Texans."

FOR ADDITIONAL INFORMATION:
Public Affairs Officer
U.S. Army Engineer District
P.O. Box 17300
Fort Worth, TX 76102

Reservoir Manager
San Angelo Lake
P.O. Box 3085
San Angelo, TX 76903

SAM RAYBURN RESERVOIR

For years, Sam Rayburn Reservoir, an impoundment of the Angelina River in east Texas, has been noted nationwide for its excellent bass fishing. Anglers came from as far away as Alaska to fish its waters, and seldom did anyone go away empty-handed. The success stories at Sam Rayburn became almost legendary. Today this remarkable lake just a few miles west of the Louisiana-Texas border is still one of the most popular in the Lone Star State. A vast reservoir covering 114,000 acres, it was named after the late speaker of the House, Sam Rayburn of Texas. It was originally known as the McGee Bend Dam and Reservoir, but by special act of Congress in 1963 the name was changed to honor Rayburn, a champion of soil and water conservation who helped make the long dream of an almost inexhaustible supply of inland water here a reality. During dedication ceremonies held in 1965, President Lyndon B. Johnson paid tribute to Rayburn, saying "he was my teacher and counselor." The dedication plaque is mounted in Texas red granite stone in the overlook circle west of the powerhouse on FM 255. Speaker Sam rosebushes are planted at the flag pole near the dam. The reservoir is partially bordered by the Angelina National Forest in the heart of Texas' Big Thicket country. Recreation facilities are operated by the Corps of Engineers, the U.S. Forest Service, and Nagadoches County.

HOW TO GET THERE: From Jasper, drive northwest for 15 mi. on SR 63, then northeast on FM 255 for about 5 mi. to dam.

FISHING: Largemouth bass is the king of the fishing world here, but anglers may also find smallmouth, bream, crappie, channel catfish, and perch. Both bank fishing and boat fishing are popular. Stands of flooded timber in some parts of the lake normally are the best spots in which to find largemouth, and some weighing more than eight pounds have been taken. Hand-size bluegill are also found here, as well as crappie the size of a saucer. For many years, Sam Rayburn has been the fisherman's dream. It is best, however, if you really want fishing success, to hire a guide. Many are available in the area. The lake is so vast that you can waste a lot of time if you're on a limited schedule just trying to find the fish and familiarizing yourself with the lake. Wind conditions are the main detriment to successful fishing here; often it's impossible to get out on the lake because of winds in excess of 25 mph. Two fishing barges on the lake, as well as an air-conditioned and heated fishing pier. Around 20 launching ramps; lakeside marinas, rental boats, supplies. Toledo Bend Reservoir, a few miles to the east, is another hot fishing spot; it and Rayburn are often rated among the top five bass lakes in the country by experts. Rayburn fish run a little larger than Toledo Bend, but Rayburn is an older lake.

HUNTING: Excellent waterfowl hunting is available on Sam Rayburn, particularly in the timbered areas. Wood duck, mallards, and several other species winter in the area. Squirrel and rabbit provide fun hunting on the surrounding areas and along the waterfront; quail and dove are also available. Deer are found in spots, but are not plentiful. Hunting opportunities are excellent in Angelina National Forest, which adjoins the reservoir.

CAMPING: Almost 200 Corps of Engineers campsites, both primitive and developed. Facilities include restrooms, drinking water, showers, tables, grills, dump station, electrical hookups, and launching ramps. Additional lakeside campgrounds are provided by the U.S. Forest Service, Nagadoches County, and private concessionaires. A KOA campground is located right on the lake, too, offering excellent camping facilities with full hookups.

OTHER ACTIVITIES: Swimming, water skiing, boating; around 20 launching ramps, lakeside marinas, rental boats. Sailboats are seen more and more often here. During the late summer when the water clears, scuba diving and snorkeling are popular. The area is well known for rockhounding. A vast variety of native flora, including orchids and insect-eating plants, as well as scores of wildlife species, provide lots of opportunity for nature study. Many scenic drives and hiking trails in area. Golf courses, tennis courts, and swimming pools are available in Lufkin, Nagadoches, or San Augustine.

AREA ATTRACTIONS: Only a few miles away from Sam Rayburn is the Alabama-Coushatta Indian Reservation, where visitors may take tours of the Big Thicket National Preserve, watch alligator wrestling, or just visit with the Indians, see their way of life, sample their foods, and purchase their craft items. Nagadoches, a few miles to the northwest, is an important historical town with many antique shops, historical buildings, a museum, and four Indian mounds within the city limits. Near Pineland, the headquarters mill of one of the area's largest lumber companies may be toured on weekdays. Sawmill Town U.S.A. at Newton offers stagecoach, buggy, and horseback rides.

FOR ADDITIONAL INFORMATION:

Public Affairs Officer
U.S. Army Engineer District, Fort Worth
P.O. Box 17300
Fort Worth, TX 76102

Project Engineer
Sam Rayburn Reservoir
Route 3, Box 320
Jasper, TX 75951

Forest Supervisor
National Forests in Texas
P.O. Box 969
Lufkin, TX 75901

TOWN BLUFF LAKE

Not far from the Big Thicket National Preserve in the East Texas Piney Woods is remarkable Town Bluff Lake, where the Angelina and Neches rivers meet. Officially named B. A. Steinhagen Lake (the dam impounding it is named Town Bluff), this small reservoir covers some 9,000 acres. Four national forests and the state's only Indian reservation, all in the same area, offer additional opportunities for recreation. Woodville and Jasper are the largest nearby towns, but this lake is also an easy drive from Houston. Public facilities on the lake include rental cabins, which are operated by the Texas Parks and Wildlife Commission.

HOW TO GET THERE: From Jasper, drive west on U.S. 190 for 15 mi., then 3 mi. south on FM 92.

FISHING: Largemouth bass, bream, red ear sunfish, crappie, and white bass inhabit these waters. This is not a deep lake, so medium to surface plugs normally work best for the big bass, live bait for other species. Lakeside marina, rental boats, supplies; around a dozen launching ramps. At Walnut Ridge Park, operated by the state, is a fishing barge.

HUNTING: Some waterfowl hunting in the late autumn, primarily ducks. Deer are found in the area, although they're not plentiful. Small game such as rabbit, squirrel, and certain game birds, quail and dove in particular, are abundant.

CAMPING: Around 90 Corps of Engineers campsites, both primitive and developed. Facilities include restrooms, drinking water, showers, tables, grills, dump station, electrical hookups, and launching ramps. Nearly 250 more lakeside sites are maintained by the Texas Parks and Wildlife Commission. Several private campgrounds are also located within 20 miles of the lake itself.

OTHER ACTIVITIES: Swimming, water skiing, boating; lakeside marina, rental boats, about a dozen launching ramps. Hiking, picnicking, and rockhounding are all worthwhile pastimes here. Portions of the Neches River offer picturesque canoeing. A drive or hike through the Angelina, Davy Crockett, Sabine, and Sam Houston national forests nearby provides an opportunity for visitors to study the vegetation and wildlife of the area.

AREA ATTRACTIONS: The Big Thicket National Preserve, one of America's newest national parks, is located just west of Woodville and extends southward for many miles toward Beaumont; called "the biological crossroads of North America" because of its unique plant community contrasts, the preserve may be explored by foot or canoe. The Alabama-Coushatta Indian Reservation is also just west of Woodville; visitors may watch tribal dances, eat authentic Indian foods at a restaurant operated by tribal members, see a Living Indian Village, purchase arts and crafts items, or take a guided tour of the Big Thicket on a train or open-air bus. At Woodville itself are the Big Thicket Gardens, which provide an educational look at native plant life. The Big Thicket Association Museum at nearby Saratoga is well worth visiting. It's near here, too, that travelers can drive down the Ghost Road, if they have the courage; the Big Thicket Association folks at the museum can tell visitors all about it and how to reach it.

FOR ADDITIONAL INFORMATION:

Public Affairs Officer
U.S. Army Engineer District, Fort Worth
P.O. Box 17300
Fort Worth, TX 76102

Reservoir Manager
B. A. Steinhagen Lake—Town Bluff Dam
Route 1, Box 82
Woodville, TX 75979

Forest Supervisor
National Forests in Texas
Box 969
Lufkin, TX 75901

WHITNEY LAKE

Just south of the megalopolis of Dallas—Fort Worth in the rolling hills of Texas is Whitney Lake on the Brazos River, an attractive recreational gem with something for virtually everyone. It's a boater's and water skier's paradise, but the fishing is excellent during some seasons here, too. There are meadows of wild flowers in the spring, forested wildlife and natural areas, picnic tables along the waterfront, and a simple scenic beauty seldom found in this part of Texas. Many people consider Whitney the loveliest lake in the Lone Star State; with over 4 million annual visitors, it's certainly one of the most popular. In fact, it currently ranks as one of the 15 most popular Corps of Engineers reservoirs in the entire country. The 50,000-acre lake attracts considerable wildlife, including shore birds, gulls, waterfowl, and upland game birds in the surrounding woodlands and meadows. Modern lodges and resorts, as well as rustic fishing camps, complement family campgrounds here. A paved airstrip is available for public use. Facilities are operated by the Corps, the Texas Parks and Wildlife Department, and private concessionaires.

HOW TO GET THERE: From Whitney, head southwest for 5 mi. on SR 22.

FISHING: During the spring and fall months, this lake provides some excellent largemouth bass fishing. Some eight-pounders have been taken from these waters, and biologists claim there are larger ones yet to be had. Bream and crappie are also available. Several areas of the lake are zoned for slow speeds of five mph, which make ideal conditions for the fisherman. A good many acres adjacent to the reservoir are being preserved as natural areas or public hunting lands, thereby helping to create better water quality and fishing along the shoreline. Private concessionaires offer two fishing barges, and there are several fish cleaning shelters around the lake. More than 15 launching ramps; lakeside marinas, rental boats, supplies.

HUNTING: Several thousand acres surrounding the lake have been set aside as public hunting areas. Deer, rabbit, squirrel and several types of game birds are the primary quarry.

CAMPING: Over 100 Corps of Engineers campsites, both primitive and developed. Facilities include restrooms, drinking water, showers, tables, grills, dump stations, electrical hookups and launching ramps. Lake Whitney State Park, which borders the shoreline, also has some 100 modern sites. Private campgrounds in area.

OTHER ACTIVITIES: Swimming, water skiing, boating; more than 15 launching ramps, lakeside marinas, rental boats. Some of the remote creeks flowing into the lake, as well as the Brazos River, are excellent canoe streams. Three natural areas and three wildlife areas which edge the reservoir provide the opportunity for nature study.

AREA ATTRACTIONS: Reconstructed Fort Graham and some still-visible scars of the Old Chisholm Trail are near the reservoir. Not far from Glen Rose, in the bed of the Paluxy River, a 26-inch imprint marks the spot where a 30-ton brontosaurus went strolling before the dawn of civilization; the rock formations here have yielded many geological and paleontological treasures, and the local historical society offers guided tours of the dinosaur tracks area. Dinosaur tracks may also be seen at Dinosaur Valley State Park north of Meridian. The Hill County Court House at Hillsboro is well worth a look as possibly the most controversial court house ever in appearance; at the same time the old *Saturday Evening Post* was denouncing it as "a monstrosity," *Harper's* was proclaiming that it was "like an outstanding cathedral." Not far to the north of Whitney Lake is the metropolitan complex of Dallas—Fort Worth with numerous attractions; contact the chambers of commerce in these cities for literature on what they have to offer. Fort Fisher in Waco is a museum of Texas Rangers relics and lore which also houses the Homer Garrison Memorial Museum of guns, weapons, and mementos of past rangers. Near Corsicana is Pioneer Village, a restored log cabin community which offers a glimpse into the century-old lifestyle of this area.

FOR ADDITIONAL INFORMATION:
Public Affairs Officer
U.S. Army Engineer District, Fort Worth
P.O. Box 17300
Fort Worth, TX 76102

Project Engineer
Whitney Project Office
P.O. Box 38

Laguna Park Rural Station
Clifton, TX 76634

WRIGHT PATMAN LAKE

Formerly this lake was known as Lake Texarkana, but in December of 1973 the president signed a bill officially designating the project Wright Patman Dam and Lake in honor of Congressman Patman of the First Congressional District of Texas. The dam, located just nine miles southwest of the city of Texarkana, backs up the waters of the Sulphur River to form a reservoir of more than 20,000 acres. Wright Patman Lake is one of the nation's greatest water playgrounds, providing something for everyone interested in the great outdoors. Numerous facilities conveniently located around the lake offer opportunity for relaxation and unsurpassed recreation, and the climate is mild enough so that many of the facilities get year-round use. Fly-in visitors will find a light plane airstrip here.

HOW TO GET THERE: From Texarkana, take SR 59 south for 8 mi.

FISHING: Fishing is not as great as at some other impoundments in east Texas, of course, for this is an older lake past peak production; but it still offers some very high-quality largemouth bass, crappie, and bluegill fishing. There are also some good-sized catfish to be caught here. In shallow areas in the spring, bowfishing for gar is a fun pastime. More than 20 launching ramps; lakeside marinas, rental boats, supplies.

HUNTING: Hunting for bushy-tailed squirrel, rabbit, white-tailed deer, and waterfowl provides the sportsman with many hours of seasonal enjoyment. Thousands of acres of reservoir hunting land are heavily populated by mallard, pintail, and canvasback ducks.

CAMPING: Over 275 Corps of Engineers campsites, both primitive and developed. Facilities include restrooms, drinking water, showers, tables, grills, dump stations, electrical hookups, and launching ramps. Atlanta State Park has over 50 more modern sites, and several private concessionaires also maintain lakeside campgrounds.

OTHER ACTIVITIES: Swimming, water skiing, boating; more than 50 launching ramps, lakeside marinas, rental boats. Visitors also enjoy hiking, picnicking, trailbike riding, and bird-watching. Nature walks are available. Golf courses are located nearby, as are riding stables.

AREA ATTRACTIONS: Texarkana is a most interesting city in itself; it has two of virtually everything—one for Texas and one for Arkansas—except the U.S. Post Office. There are two mayors, two fire chiefs, and two school systems, but the state line runs squarely through the center of the post office. Residents claim the post office is the only one in America located in two states. The Red River Army Depot west of Texarkana, which stores and repairs many types of arms and equipment, may be toured, but call in advance. Texans call much of this area "sweet potato country," and you'll see why if you drive about a bit. You'll also find some interesting little towns, such as Redwater and Sulphur Springs, filled with genuine Texas hospitality and maybe a tall tale or two as well if you'll pause to chat with some of the old-timers. There are 70 Indian mounds in the vicinity of Wright Patman Lake.

FOR ADDITIONAL INFORMATION:

Public Affairs Officer
U.S. Army Engineer District, New Orleans
P.O. Box 60267
New Orleans, LA 70160

Resource Manager
Wright Patman Lake
P.O. Box 1817
Texarkana, TX 75501

ADDITIONAL CORPS OF ENGINEERS LAKES*

Addicks Dam (in northwest Houston); Bardwell Lake (from Ennis, 4.5 mi. S on county paved road and SR 34); Belton Lake (from Belton, 3 mi. N on SR 317, 1 mi. NW on FM 2271); Benbrook Lake (from Benbrook, 1 mi. S on

*Unless designated as follows, each project has restrooms, drinking water, and developed campsites. † = no restrooms; ‡ = no drinking water; § = no developed campsites.

U.S. 377, 2.5 mi. E on county road); Canyon Lake (from New Braunfels, 15 mi. NW on FM 306); Hords Creek Lake (from Coleman, 8 mi. W on FM 53, S on county road); Lavon Lake (from Wylie, 3 mi. NE on SR 78, 1 mi. N on county road); Navarro Mills Lake (from Corsicana, 20 mi. SW on SR 31, 1 mi. N on FM 667); Pat Mayse Lake (from Paris, 15 mi. N on U.S. 271, 3 mi. W on FM 906); Proctor Lake (from Comanche, 5 mi. E on U.S. 377, 2 mi. N on county road); Somerville Lake (from Somerville, 1 mi. W on SR 36); Stillhouse Hollow Lake (from Belton, 5 mi. SW on U.S. 190, 4 mi. left on FM 1670); Waco Lake§ (from Waco, 2 mi. NW on FM 1637, left for 1.5 mi.)

VERMONT

BALL MOUNTAIN RESERVOIR

Tiny Ball Mountain Reservoir, with a shoreline of only two miles, extends up the narrow West River Valley in south central Vermont. The heavily wooded reservoir lands, flanked by Ball Mountain and Shatterack Mountain, lie in a highly scenic area and will be maintained in their natural state. Just north and west of Ball Mountain Reservoir is Green Mountain National Forest with additional recreational opportunities. Facilities at Ball Mountain are somewhat limited, but the lake draws many sightseers and sportsmen.

HOW TO GET THERE: From Brattleboro, take I-91 north to SR 30, then continue north for about 35 mi. to Jamaica and follow signs 2 mi. north to dam.

FISHING: Trout fishing is a way of life in Vermont, but anglers will also find smallmouth bass, walleye, and northern pike.

HUNTING: White-tailed deer are the chief big game species; so numerous are they that overpopulation has sometimes been a problem. Windham County, in which Ball Mountain Reservoir lies, is one of the top whitetail areas in the state. Rabbit, squirrel, grouse, woodcock, and pheasant are also hunted. Fair waterfowl hunting along the river when floated in canoes.

CAMPING: Over 50 Corps of Engineers primitive campsites; facilities include restrooms, drinking water, tables, and grills. Jamaica State Park, on the banks of the West River nearby, offers nearly 50 developed campsites with pay showers and a dump station. There are many private campgrounds in the area.

OTHER ACTIVITIES: Lake and river swimming. The West River reach between Ball Mountain Reservoir and Townshend Dam is one of the East's major whitewater canoeing and kayaking centers. Many visitors combine sightseeing with a picnic lunch. Winter activities are snowmobiling and cross-country and downhill skiing; several ski resorts are within a half-hour's drive. With mountains all around, few people can resist climbing or hiking one. Scenic drives in every direction. Rental horses nearby. The Appala-chian Trail and Long Trail follow the same route a few miles west of the reservoir.

AREA ATTRACTIONS: Bellows Falls has interesting Indian carvings on rocks along the river, as well as Steamtown U.S.A., where visitors can ride a 70-year-old steam excursion train. The original Vermont Country Store near Weston displays wares for homes of the 1890s. North of Ludlow the public is invited to visit the century-old Crowley Cheese Factory, where cheese is still made by hand. Experienced mountain drivers will want to follow the 6-mile road to the top of 3,816-foot Mount Equinox near Manchester; but all cars should be in top-notch condition before attempting the climb. The view from the top of Bromley Mountain near Manchester, accessible by chair lift, sweeps five states. Another Manchester attraction is the Southern Vermont Art Center, which offers exhibits and summer concerts. More summer concerts are featured at Marlboro's Music Festival.

FOR ADDITIONAL INFORMATION:

Public Affairs Officer
U.S. Army Engineer Division, New England
424 Trapelo Road
Waltham, MA 02154

Jamaica State Park
Jamaica, VT 05343

Forest Supervisor
Green Mountain National Forest
151 West St.
Rutland, VT 05701

NORTH HARTLAND LAKE

From its mountain origin, the Ottauquechee River rushes eastward toward its junction with the mighty Connecticut, until it reaches Quechee Gorge at the upper end of North Hartland Reservoir. Here it narrows, then plunges downward in a 155-foot drop to pass through the sheer-faced gorge, one of the Green Mountain State's outstanding natural spectacles. Just beyond this wild passage a valley bordered by gently sloping hills enables a 220-acre reservoir to form behind North Hartland Dam in east central Vermont. The noted nineteenth-century conservationist, George Perkins

Horseback riding is a favorite pastime at many Corps of Engineers sites. (Corps of Engineers photo)

Marsh, was a native of nearby Woodstock; and the Corps of Engineers has named the conservation pool Marsh Conservation Lake in his honor. Recreational facilities are managed by the Corps and by the state of Vermont.

HOW TO GET THERE: From White River Junction, take SR 5 south for 5 mi.

FISHING: Trout are probably the most popular catches, but anglers will find both cold- and warm-water species here; among them are largemouth and smallmouth bass, bream, catfish, bluegill, walleye, and northern pike. The Corps of Engineers maintains a launching ramp; rental boats.

HUNTING: The environs of Windsor County, in which North Hartland Reservoir lies, are known for their productive deer hunting. Cottontail rabbit and snowshoe hare are both found here, but cottontails are more abundant. Also pheasant, grouse, wood-cock, and squirrel. Some good duck hunting along the eastern shore of the Connecticut River.

CAMPING: Over 25 Corps of Engineers developed campsites; facilities include restrooms, drinking water, showers, tables, grills, and launching ramp. Quechee Gorge State Park offers another 30 modern campsites with pay showers and dump station. Private camp-grounds in area.

OTHER ACTIVITIES: Boating; launching ramp, rental boats. A scenic walkway leads from the dam to the Ottauquechee River Falls. Picnicking along reser-voir shores. The Corps hopes to develop this reservoir into a major recreation area, and facilities are being added all the time. A swimming beach is planned for the future. Woodstock has a public golf course. Snow-mobiling allowed when weather conditions permit. Some of Vermont's famous ski resorts are not far

away. Barrel stave skiing is a popular sport at Kill-ington Ski Area near Rutland.

AREA ATTRACTIONS: The American Precision Museum in Windsor, depicting the history of the in-dustrial revolution, is housed in a building which is designated as a National Historic Site. Woodstock, where three covered bridges span the Ottauquechee River, is known as one of New England's most attrac-tive villages. Calvin Coolidge was born, grew up, and took the presidential oath (upon President Harding's death) in Plymouth. There are several granite quarries around Rutland. A huge marble exhibit and the interesting Wilson Castle are at Proctor.

FOR ADDITIONAL INFORMATION:
Public Affairs Officer
U.S. Army Engineer Division, New England
424 Trapelo Rd.
Waltham, MA 02154
Quechee Gorge State Park
R.F.D.
White River Jct., VT 05001

ADDITIONAL CORPS OF ENGINEERS LAKES*

North Springfield Lake, ‡, § (from Hartford, N on I-91 to Vt. exit 7, N on SR 106 to North Springfield, turn right on Maple St., follow signs to office on Reservoir Rd.); Townshend Lake § (from Brattleboro, N on SR 30 about 20

*Unless designated as follows, each project has restrooms, drinking water, and developed campsites. † = no restrooms; ‡ = no drinking water; § = no developed campsites.

Swimming opportunities are endless at Corps of Engineers lakes. (Corps of Engineers photo)

mi.); Union Village Dam § (from White River Junction, N on U.S. 5 to Pompanoosuc, left on SR 132); Waterbury Lake (from Montpelier, 11 mi. W on I-89 to Waterbury, N on SR 100); Wrightsville Lake ‡, § (on SR 12, just N of Montpelier)

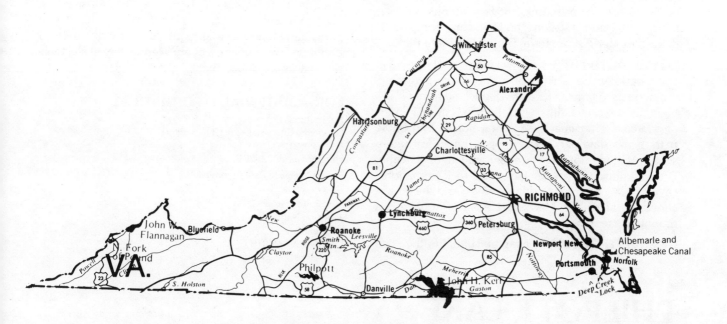

VIRGINIA

JOHN W. FLANNAGAN RESERVOIR

Along the narrow valleys of the Pound and Cranesnest rivers in southwest Virginia, this small, serpentine reservoir seeks its way through the mountains and forests which surround it. This is a land rich in legend and lore, of caves and Shawnee powwows, buried treasure, Daniel Boone, and the Hatfields and McCoys. Jefferson National Forest adjoins reservoir lands, and just to the north is the unique Breaks Interstate Park, coadministered by Virginia and neighboring Kentucky. The combination of lake, forest, and park provides an unlimited amount of recreational opportunities. Both Virginia and Kentucky have lodges at Breaks Interstate, and Kentucky offers housekeeping cottages as well.

HOW TO GET THERE: From Haysi, VA, head west on SR 83 to SR 614 and SR 739. The total route is 7 mi. long.

FISHING: Trout fishing at its best in lake, feeder streams, and nearby mountain brooks; also large-mouth, smallmouth, and spotted bass, pickerel, walleye, muskie, and panfish. Nearly half a dozen Corps of Engineers launching ramps; lakeside marina, rental boats, supplies.

HUNTING: Game includes deer, wild turkey, squirrel,

rabbit, grouse, quail, and dove. Hunting in Jefferson National Forest and on nearby state wildlife management area.

CAMPING: Nearly 100 Corps of Engineers campsites at reservoir's edge, both primitive and developed. Facilities include restrooms, showers, drinking water, tables, grills, dump stations, and launching ramps. Additional modern campsites at Breaks Interstate Park.

OTHER ACTIVITIES: Lake swimming, public pool in state park, water skiing, pleasure boating; marina, rental boats, nearly half a dozen launching ramps. Hiking trails and scenic drives in Jefferson National Forest and Breaks Interstate Park; rental horses and bridle path at the state park. Lots of opportunity for nature study. Varied mountain scenery, much of it within Breaks Interstate Park, includes odd rock formations, caves, balanced rocks, hidden springs, and the largest canyon east of the Mississippi River. Canoeing on mountain streams nearby.

AREA ATTRACTIONS: Natural Tunnel in the state park of the same name, as high as a 10-story building, is a scenic wonder. The history, culture, and industry of this region is exhibited at the Southwest Virginia Museum in Big Stone Gap. In the same city, *Trail of the Lonesome Pine,* an outdoor musical drama, is presented during July and August. The unusual Bee Hive coke ovens are still used in Wise County; they're particularly impressive at night. All that old-fashioned bartering which goes on every Wednesday at Norton's stockyards is a public event.

FOR ADDITIONAL INFORMATION:

Public Affairs Officer
U.S. Army Engineer District, Huntington
P.O. Box 2127
Huntington, WV 25721

Breaks Interstate Park Commission
Breaks, VA 24607

Forest Supervisor
Jefferson National Forest
920 Jefferson St.
Roanoke, VA 24016

PHILPOTT LAKE

Beyond the wooded foothills which embrace Philpott Lake's 100-mile shoreline can be seen the crest of the Blue Ridge Mountains, a stunning backdrop for this many-fingered lake. Fairy Stone State Park, which partially borders Philpott Lake, has its own 168-acre lake adjoining this Corps of Engineers reservoir in south central Virginia. Together, the state and the Corps offer a full spectrum of recreational activities.

HOW TO GET THERE: From Roanoke, head south on U.S. 220 for 47 mi., turn west on SR 57, then north on County Rd. 904.

FISHING: The Smith River, which feeds the reservoir, is a noted trout stream. Largemouth and smallmouth bass, crappie, and bream are principal fish in the reservoir itself. Walleye, coho salmon, and channel catfish have also been stocked here. Nearly a dozen launching ramps; lakeside marina, rental boats, supplies.

HUNTING: Grouse and deer provide some of the best hunting here, and there's a season for wild turkey. Other small game includes rabbit, squirrel, and quail. Fairystone Farms Wildlife Management Area allows in-season hunting on land adjoining it, and hunting is allowed at designated spots around the reservoir.

CAMPING: Over 200 Corps of Engineers campsites, both developed and primitive. Facilities include restrooms, drinking water, showers, tables, grills, firewood, disposal station, and launching ramps. Several campsites accessible by boat only; one campground is on Deer Island. Group camping area near the dam. Fairy Stone State Park has rental cabins.

OTHER ACTIVITIES: Swimming, water skiing, skin and scuba diving, boating; lakeside marina, rental boats, nearly a dozen launching ramps. Canoeing is becoming increasingly popular. Several picnicking spots around reservoir; some accessible by boat only. Lucky "fairy stones" are found in and around the state park; a nature center features their legend. Horses may be rented in the state park for rides along the hillsides. An old iron mining village may be reached by a footpath. The state park also features several hiking and biking trails.

AREA ATTRACTIONS: Blue Ridge Parkway, one of the nation's top scenic drives, is about 10 miles to the west. Just beyond it, the vast Jefferson National Forest is waiting to be explored; the Appalachian Trail passes through a portion of it. The Booker T. Washington National Monument near Roanoke is a working plantation well worth seeing. At Meadows of Dan a side wheel paddleboat chugs about over a short course among the mountaintops. The public is welcome on fall weekdays to see the tobacco auctions in Danville. Mill Mountain reaches upward 2,000 feet; it's com-pletely within Roanoke's city limits, and visitors can drive to its summit.

FOR ADDITIONAL INFORMATION:

Public Affairs Officer
U.S. Army Engineer District, Wilmington
P.O. Box 1890
Wilmington, NC 28401

Reservoir Manager
Philpott Lake
Rte. 6, Box 140
Bassett, VA 24055

Fairy Stone State Park
Stuart, VA 24171

ADDITIONAL CORPS OF ENGINEERS LAKES*

Locks on the Atlantic Intracoastal Waterway: Great Bridge§ (in Chesapeake, on Albemarle and Chesapeake Canal, 3.5 mi. S of I-64 on SR 168) and Deep Creek § (in Chesapeake, on Dismal Swamp Canal, 2 mi. S of I-64 on U.S. 17); North Fork of Pound Lake (from Pound, 1 mi. SW on SR 630)

WASHINGTON

ICE HARBOR LOCK AND DAM (LAKE SACAJAWEA)

Located near the confluence of the Snake and Columbia rivers in southeastern Washington, long, narrow Lake Sacajawea lies in country traversed by the Lewis and Clark Northwest Expedition. To the east is the Evergreen State's wheat country, and the rich harvest is shipped by barge through Ice Harbor Lock and on to the Columbia River. Visitors to the dam can watch Pacific salmon and steelhead trout en route to their spawning grounds, then pause to enjoy the recreational facilities provided by the Corps of Engineers.

HOW TO GET THERE: From Burbank, head east for 8 mi. on SR 124.

FISHING: Steelhead trout are the most highly prized game fish in the state, and they're abundant here during spawning runs. The Snake River also is tops for smallmouth bass and channel catfish. Lakeside marina, rental boats, nearly half a dozen launching ramps.

HUNTING: Duck and goose hunting is excellent. Deer, pheasants, grouse, and quail also call this region home.

---

*Unless designated as follows, each project has restrooms, drinking water, and developed campsites. † = no restrooms; ‡ = no drinking water; § = no developed campsites.

CAMPING: Around 100 Corps of Engineers developed campsites. Facilities include some full hookups, restrooms, showers, drinking water, tables, grills, dump stations, playground, firewood, and launching ramps. Private campgrounds in vicinity provide additional sites.

OTHER ACTIVITIES: Swimming at four beaches, water skiing, boating; lakeside marina, rental boats. Hiking along nature trails, picnicking along shoreline. Scenic drives along Snake River.

AREA ATTRACTIONS: The McNary National Wildlife Refuge is near Burbank. In Richland the public may visit the Hanford Science Center and explore the mysteries of atomic energy. Sacajawea State Park and Museum southeast of Pasco bring to life the adventures of the Lewis and Clark Expedition and honor the Shoshone Indian girl who served as its guide. The Whitman Mission National Historic Site relives the history of an Indian school and massacre.

FOR ADDITIONAL INFORMATION:
Public Affairs Officer
U.S. Army Engineer District, Walla Walla
Bldg. 602, City-County Airport
Walla Walla, WA 99362

ADDITIONAL CORPS OF ENGINEERS LAKES*

Chief Joseph Dam, or Lake Rufus Woods (from Chelan, 32 mi. N on SR 97, 9 mi. E on SR 17); Lake Washington Ship Canal § (in the Ballard District of Seattle); Little Goose Lock and Dam (from Starbuck, 8 mi. NE); Lower Granite Lock and Dam †, ‡, § (from Pomeroy, 20 mi. NE, follow signs); Lower Monumental Lock and Dam (from Kahlotus, 6 mi. S); Mill Creek Lake § (from Walla Walla, 2 mi. E); Mud Mountain Dam § (from Enumclaw, 8 mi. E on SR 410); Wynoochee Lake § (from Montesano, 35 mi. N on county and Forest Service roads)

---

*Unless designated as follows, each project has restrooms, drinking water, and developed campsites. † = no restrooms; ‡ = no drinking water; § = no developed campsites.

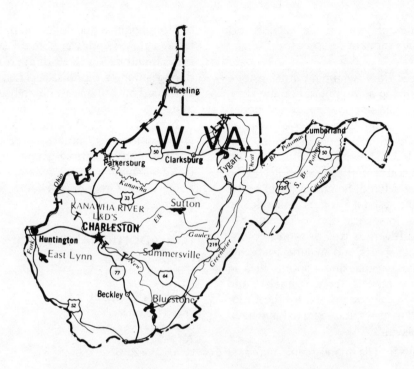

WEST VIRGINIA

BLUESTONE LAKE

Majestic mountains, dense forests, and the magnificent New River, calmed temporarily by Bluestone Lake, create a public playground which lures over a million people annually. The reservoir and the river which creates it lie along a valley which offers some of the state's loveliest scenery. About 45 miles below the dam is the rugged New River Gorge, where the roar of the water may sometimes be heard a mile away. Oddly enough, the New River is rated by some geologists as the oldest river in the world. Located in southeastern West Virginia near the Virginia border, Bluestone Lake represents a cooperative effort of the Corps of Engineers, which maintains the project, and the state, which administers most of the recreation facilities, including vacation cabins.

HOW TO GET THERE: From Hinton, take SR 20 south for 1 mi. and west for another mi.; turn south on SR 3 for about 1 mi. to the dam.

FISHING: Some good catches of crappie, smallmouth bass, and channel and flathead catfish have been made in the reservoir, while the tailwaters are noted for muskie. Brook, brown, rainbow, and golden trout provide thrills for New River anglers. Over half a dozen launching ramps; lakeside marina, rental boats, supplies.

HUNTING: A wide variety of game includes squirrel, rabbit, ruffed grouse, quail, dove, and woodcock. Deer and turkey hunting is light because of restrictions. Bluestone Public Hunting Area is near the lake.

CAMPING: Over 80 developed campsites at Bluestone State Park along the reservoir's edge; facilities include restrooms, drinking water, showers, tables, grills, firewood, playground, dump station, and launching ramps. There are an additional 400 primitive sites in Bluestone Public Hunting Area. Pipestem

157

Resort State Park, though not on the lake, is just minutes away with more modern campsites.

OTHER ACTIVITIES: Pool swimming at Bluestone State Park, lake swimming at the public hunting area, water skiing, boating; lakeside marina, rental boats, over half a dozen launching ramps. Picnic areas all around reservoir. Excellent rockhounding; many artifacts remain from an Indian town which once stood on project lands. Game courts and playground. Nearby Pipestem Resort is considered the crown jewel of the Mountain State's park system; it offers trail rides to wilderness areas on horseback and wildwater raft trips down Bluestone River, as well as tennis courts, two golf courses, winter sports, and an aerial tramway. Numerous hiking and backpacking opportunities in area. The New River attracts more canoeists and rafters than any river in the country except the Colorado. Shunpiking and nature study seem to have been invented for country like this.

AREA ATTRACTIONS: The many scenic wonders of Monongahela and Jefferson national forests are not far away. At Grandview State Park visitors can catch some breathtaking views of the New River Gorge and, during summer months, see two acclaimed outdoor dramas. Tours of a real coal mine with working equipment and displays are conducted by old-time miners in New River City Park near Beckley. Railroad buffs will want to take a short trip on Bluefield's Ridge Runner Railroad, a narrow-gauge train; more dedicated devotees may wish to make a day-long excursion aboard the Greenbrier Scenic Railroad from Ronceverte.

FOR ADDITIONAL INFORMATION:
Public Affairs Officer
U.S. Army Engineer District, Huntington
P.O. Box 2127
Huntington, WV 25721
Project Superintendent
Bluestone Lake
701 Miller Ave.
Hinton, WV 25951

TYGART LAKE

Tucked amid the mountains and forests of northeastern West Viriginia, reaching for 6 miles up the Tygart River Valley, is a scenic 1,700-acre lake which attracts both summer and winter sports enthusiasts. In this land of diversity visitors will find spectacular scenery, arts and crafts practiced since our country's beginnings, land where time stands still, and nearly impenetrable wilderness, complemented by the most modern of resorts. Recreation facilities along the reservoir are administered primarily by the state. Tygart State Park offers a beautiful guest lodge and rental cabins.

HOW TO GET THERE: From Grafton, follow a marked road 3 mi. south to the dam.

FISHING: Hundreds of rushing mountain streams provide excellent trout fishing. The reservoir produces some fine largemouth bass, walleye, muskie, crappie, bluegill, catfish, and northern pike; and there are some good smallmouth streams nearby. Ice fishing in winter. Three launching ramps; lakeside marina, rental boats, supplies.

HUNTING: The state's Pleasant Creek Hunting Area adjoins the reservoir, and Monongahela National Forest is less than 30 miles away. White-tailed deer, wild turkey, quail, rabbit, squirrel, and ruffed grouse roam the woodlands. The Doe Run Waterfowl Impoundment on the lake attracts ducks.

CAMPING: Nearly 35 campsites, both primitive and modern, at Tygart Lake State Park; another 40 developed sites at Grafton Municipal Park near the dam; and over 30 sites for primitive camping in the Pleasant Creek Hunting Area. Facilities include restrooms, drinking water, tables, firewood, playground, game courts, laundromat, dump stations, and launching ramps. Private campgrounds nearby.

OTHER ACTIVITIES: Lake swimming, water skiing, skin and scuba diving, pleasure boating of all types; lakeside marina, rental boats, and three launching ramps. Plenty of picnic spots. Lots of area in which to explore by foot or auto. The Tygart Lake Country Club and Golf Course are near the lakeshore, and there's a rifle range on one of the tributary creeks. Good rockhounding in area. Excellent canoeing on the Tygart River and on mountain streams not far away. Canaan Valley State Park, a little over an hour's drive away, is a major snow skiing area. The entire region is ideal for shunpiking.

AREA ATTRACTIONS: Valley Falls State Park is a scenic treasure, featuring the swift falls of the Tygart River. A majestic stand of timber at Cathedral State Park, designated a National Natural History Landmark, may be explored by hiking trails. The deep canyon and lovely falls at Blackwater Falls State Park are also worth seeing. There's a pioneer homestead restoration at Watters Smith State Park. At French Creek Game Farm, operated by the state, the public may view over 50 kinds of West Virginia wildlife. The nearby Monongahela National Forest offers a myriad of stunning scenic attractions. At Grafton visitors may see a national cemetery and the Mother's Day National Shrine (the holiday was started here in 1908). The state is noted for its hand-blown glassware, and several factories in the area offer tours.

FOR ADDITIONAL INFORMATION:

Public Affairs Office
U.S. Army Engineer District, Pittsburgh
Federal Bldg., 1000 Liberty Ave.
Pittsburgh, PA 15222

Tygart Lake State Park
Attn: Park Superintendent
R.D. 1
Grafton, WV 26354

ADDITIONAL CORPS OF ENGINEERS LAKES*

East Lynn Lake (from Wayne, 12 mi. SE on SR 37); Summersville Lake (on SR 129 at Mt. Nebo, 3 mi. W of U.S. 19); Sutton Lake (1 mi. E of Sutton)

WISCONSIN

EAU GALLE RESERVOIR

Located in a scenic area with steep hills, valleys, bluffs, streams, and lakes, this 940-acre lake is a popular day-use-only recreation area. It lies on the Eau Galle River in west central Wisconsin, midway between Eau Claire, WI, and St. Paul, MN. Recreation facilities are operated by the Corps of Engineers; drinking water and restrooms are provided. There are numerous private campgrounds, motels, and restaurants in the surrounding area; and I-94 passes just north of here.

HOW TO GET THERE: From St. Paul, MN, take I-94 east to U.S. 63; turn south to junction with SR 29, then east to Spring Valley and north on dam road.

FISHING: Walleye, northern pike, smallmouth and largemouth bass, and panfish lure anglers. Two launching ramps available; nonmotorized boating only.

HUNTING: The top game in the surrounding countryside is white-tailed deer. Hunters will find grouse, partridge, woodcock, and rabbit as well. Since this region is not far from the Mississippi Flyway, there's some fine migratory waterfowl hunting.

CAMPING: None at present. Several private campgrounds in area, including a KOA at Menomonie, WI.

*These projects all have restrooms, drinking water, and developed campsites.

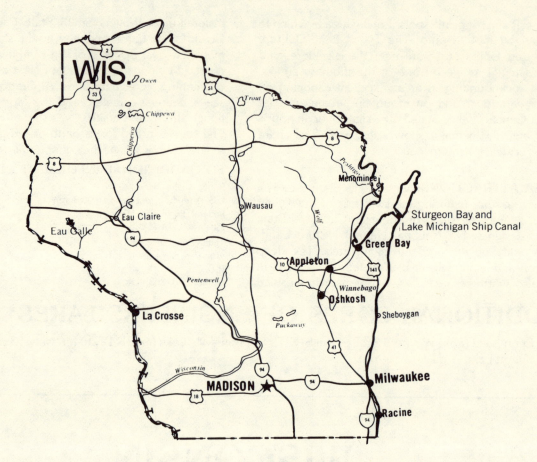

OTHER ACTIVITIES: Swimming at a sandy beach and nonmotorized boating highlight water sports. There's a change house for bathers. Along the shores are scenic picnic areas and a well-equipped children's playground. The White Pine Nature Trail, featuring much of this region's unique fauna, winds its way through a wooded glen. The St. Croix and Chippewa rivers nearby are good canoe streams. There is also the opportunity for backcountry bicycling in a state to which bicycle trails are important.

AREA ATTRACTIONS: Crystal Cave near Spring Valley may be toured. Eau Claire's Carson Park contains an old lumber camp, a one-room schoolhouse, and the Paul Bunyan Camp Museum. In nearby Chippewa Falls, Irvine Park features a native-animal zoo; there also are tennis courts and a swimming pool. Visitors are welcome at a fish hatchery near St. Croix Falls. South of the same city is Interstate State Park with its scenic gorge, the Dalles of the St. Croix River. Boat trips on the river are offered from Taylors Falls, MN. The twin cities of St. Paul and Minneapolis, MN, offer attractions too numerous to be listed.

FOR ADDITIONAL INFORMATION:
Public Affairs Officer
U.S. Army Engineer District, St. Paul
1135 U.S. Post Office and Custom House
St. Paul, MN 55101

ADDITIONAL CORPS OF ENGINEERS LAKES*

Sturgeon Bay and Lake Michigan Ship Canal* (from Milwaukee, N on U.S. 141 to Manitowoc, N on SR 42 to Sturgeon Bay, right on Memorial Dr.)

*This project has no restrooms, drinking water, or developed campsites.